W9-CHS-225

Gaining New Military Capability

An Experiment in Concept Development

John Birkler

C. Richard Neu

Glenn Kent

Prepared for the
Office of the Secretary of Defense

RAND
A NATIONAL RESOURCE
1948 ~ 1998

**National Defense
Research Institute**

PREFACE

This report focuses on the process of modernizing U.S. military forces. We believe this process requires reinvigorating concept development efforts, thinking broadly about alternatives, and pursuing concept development before decisions are made about which services, which platforms, or which technologies are best suited for accomplishing current or new military tasks.[1]

The research described here is an experiment to determine if a new and unusual approach can generate multiple ideas to better perform existing, as well as emerging, military tasks.

The report should be of interest to those involved with the modernization of U.S. military forces.

This research was conducted for the NDRI Advisory Board within the Acquisition and Technology Policy Center of RAND's National Defense Research Institute, a federally funded research and development center sponsored by the Office of the Secretary of Defense, the Joint Staff, the unified commands, and the defense agencies.

The initial results of this RAND-initiated effort were judged promising, and the NDRI Advisory Board suggested that the methodology be applied to a military task identified by United States Special Operations Command (SOCOM). We are now working closely with

---

[1]For ideas on how such a process might be institutionalized, see Paul Bracken, John Birkler, and Anna Slomovic, *Shaping and Integrating the Next Military: Organization Options for Defense Acquisition and Technology*, DB-177-OSD, Santa Monica, Calif.: RAND, 1996, pp. 24–27.

SOCOM to demonstrate the effectiveness of this process at generating concepts to gain new and potentially revolutionary capabilities.

CONTENTS

FIGURES

TABLE

SUMMARY

In the face of changing operational requirements, expanding technological opportunities, and ever more restricted budgets, simply buying "more of the same" will prove an inadequate strategy for equip-
ping the next generation of military forces. But to make effective use
of new technologies to perform changing military tasks, the military
must develop new technological and operational *concepts* for performing a wide range of military operations.

Unfortunately, current practice in the U.S. Department of Defense
(DoD) frequently works to limit systematic thinking about alternative
approaches to challenging military tasks. Ideally, broad thinking
about how specific missions might be accomplished should precede
decisions about what kinds of platforms, what classes of technologies, or which military service is most appropriate for particular
tasks. Too often, however, these steps in the process of planning
force modernization are reversed. As a result, concept development
sometimes becomes more an exercise in finding a use for a given
technology, platform, or operational method rather than in finding
the right technology or platform to perform a specific function.
Thinking about alternative options is narrowed, and competition
among alternative concepts is weakened.

Improving the force modernization process will require reinvigorating concept development efforts, reestablishing the proper emphasis
on thinking broadly about alternatives, and pursuing concept development before decisions start to be made about which services,
which platforms, or which technologies are best suited for particular
tasks. RAND has begun to experiment with a new approach to con-

cept development in the hope of providing a template for strengthening the concept development process with DoD.

CONCEPT OPTIONS GROUPS: A NEW FORUM FOR CONCEPT DEVELOPMENT

At the heart of RAND's proposal to reinvigorate the concept development process is a concept options group (COG), a small team of around 15 members working intensively over a short period of time to formulate alternative technological and operational approaches to accomplishing specific military tasks. In our view, a COG should include three kinds of members: broadly knowledgeable technologists drawn from a variety of scientific and engineering backgrounds, experienced military operators, and generalist defense analysts.

The idea behind a COG is that, by combining the expertise of technologists with the practical experience of the military operators and the broad perspective of strategic planners, it may be possible in a relatively short time to identify (but certainly not to work out in full detail) a set of technologically and operationally feasible options for performing key military tasks. Members of a COG are not chosen to represent particular interests or viewpoints. Indeed, the whole idea is to assemble a group that is able to think about alternatives for performing a mission without institutional constraints.

As an experiment, RAND convened a COG in February and March 1996 to consider two specific military tasks within the general operational context of trying to enforce a cessation of hostilities between two opposing factions. (U.S. operations in the former Yugoslavia were, of course, much on our minds.) The two specific tasks were: stopping or preventing artillery, mortar, and sniper attacks against designated targets or areas; and maintaining persistent surveillance of selected areas. We chose these tasks because we believed them likely to become increasingly important for U.S. forces, because we assessed current approaches to be inadequate, and because we perceived that minds were open to new approaches.

The participants in our COG experiment produced options for using technologies related to operational concepts that could enable U.S. forces to perform an existing military mission better, perform it differently, or gain a new capability.

The COG held three two-day meetings, with two weeks between meetings. The COG addressed five questions regarding each of the military tasks under consideration:

1. What, in very specific terms, will U.S. forces have to do to accomplish the military task being discussed?

2. What are the characteristic signatures of relevant targets or hostile activities?

3. What technologies could be used to exploit the signatures or vulnerabilities of hostile forces?

4. How could the technologies identified in discussions of the last question be embedded in practical concepts of operations?

5. How can we shape the environment to make U.S. technological and operational options more effective?

Consideration of these questions took on an iterative character, as the COG sought continually more refined answers to each.

SOME PROMISING CONCEPTS

Among the dozens of ideas raised during the COG meetings, four seemed to show particular promise. These were the subjects of detailed discussions.

UAV Operations to Support Persistent Surveillance

The COG envisioned combined operations of two kinds of unmanned aerial vehicles (UAVs) to provide persistent monitoring of the presence or movements of trucks, armored vehicles, or artillery pieces within large surveillance areas. Large UAVs orbiting continuously at high altitude would use microwave or millimeter-wave radars operating in both synthetic aperture radar (SAR) and moving target indicator (MTI) modes to detect vehicles in the open. An additional ultra-wide-band radar would provide some capability to penetrate foliage. Targets identified by high-flying UAVs would be inspected more closely by smaller UAVs operating at lower altitudes, under weather. These low-flying UAVs would carry low-light-television, imaging infrared sensors, laser radar, and/or hyperspectral sensors. Because of the risks inherent in operating at low altitudes, low-flying UAVs would be called into dangerous areas only when required to check out targets first identified by high-flyers. Both types of UAVs would carry laser designators to illuminate targets for attack.

Ground-Based, Multimode Sensor Arrays

The COG also saw considerable utility in developing small, cheap, unobtrusive multisensor packages for remote monitoring of limited geographical areas, roads, choke points, or terrain shielded from other kinds of monitoring. These packages would include acoustic, thermal, magnetic, and seismic sensors, which in combination might go a long way toward effective target recognition. Perhaps these sensors could also cue imaging sensors (television or infrared), which require too much power for continuous operation, when specific kinds of targets appear to be nearby. The COG noted that real-time monitoring is not always required in peacekeeping operations. Some suggested, therefore, that sensors might transmit information only when queried by friendly forces, thus reducing emissions, power requirements, and the likelihood that sensors might be found or disabled by unfriendly forces. Although networks of sensors reporting on specific events from multiple vantage points could be immensely valuable, few members of the COG thought full netting of sensors was likely in the next several years. They thought it a sufficient challenge to design multisensor packages that can provide useful information in isolation.

Techniques for Detecting Small Arms

The best way to defend against snipers is to prevent them from ever getting into positions to cause harm. COG discussions identified two possible techniques for limiting the ability of a sniper to move through an urban environment with a concealed weapon. At ranges of a few feet, low-power millimeter-wave radars can produce images of metallic and even some plastic weapons. Because this radiation penetrates thin, nonmetallic materials with little attenuation, imaging devices could detect weapons hidden under clothing or carried in nonmetallic bags. Imaging devices could also be placed behind thin walls (in corridors or building entrances, say) to allow clandestine scanning of persons passing through. A second, more speculative approach to detecting concealed weapons would exploit the fact that most firearms are characterized by metallic cylinders (the barrel) of roughly predictable lengths and diameters. Such cylinders will resonate and re-radiate incident radio-frequency energy at characteristic wavelengths with a different polarity from the incident radiation. Systems generating low-power radio-frequency emissions and equipped with detectors to recognize these polarity shifts can detect the presence of metallic cylinders of specific dimensions. (Commercial retailers use a similar technology today to control theft.) Such systems cannot provide images, but they should be able to detect whether anyone in a room, corridor, narrow street, etc. is carrying a metallic cylinder of suspicious dimensions.

Localizing Sources of Sniper Fire

If a sniper does manage to get into position to fire, U.S. forces will require techniques to identify quickly the location from which the attack came. COG discussions identified three techniques for accomplishing this. Multiple high-frame-rate infrared imagers might be able to observe the track of a projectile through the air, allowing trajectories to be backtracked to the source of fire. Netted acoustic sensors might similarly calculate the trajectory of projectiles moving at supersonic speeds by detecting the associated shock wave. Finally, and most speculatively, it may be possible to detect the electromagnetic pulse generated by the plasma of hot gases produced in a rifle breech during firing. COG members emphasized the importance of integrating any of these detection techniques with a system of accurate, frequently updated, and highly detailed digital maps of relevant urban areas so that trajectory data can be instantly superimposed on maps to indicate which building, for example, is sheltering a sniper.

SOME OBSERVATIONS ON THE COG PROCESS

Overall, the experimental COG seemed to function well. The group did well in formulating and discussing new ideas and concepts. Interaction within the group brought about considerable change in and refinement of views expressed early in discussions. And we believe that some useful concepts were detailed. COG discussions also illustrated clearly the different contributions of technologists, experienced military operators, and generalist defense analysts and underlined the necessity of including all three types of participants in future COGs.

Two COG areas appear to call for improvements. Without the support of a technological "backroom operation" to carry out rough calculations relevant to technologies being discussed, COG discussions sometimes lacked technical specificity. Creating a technical support staff to assist future COGs may be worthwhile. More important, we found that we had insufficient input from experienced military commanders. In planning future COG exercises, we will seek to involve a wider circle of operational military officers.

The initial results of this RAND-initiated effort were judged promising, and the NDRI Advisory Board suggested that the methodology be applied to a military task identified by the U.S. Special Operations Command (SOCOM). We are now working closely with SOCOM to demonstrate the effectiveness of this process at generating concepts to gain new, and potentially revolutionary, capabilities.

ACKNOWLEDGMENTS

This report owes much to David Gompert, RAND's National Defense Research Institute director, who supported, encouraged and gave us the freedom to explore our ideas about reinvigorating the concept development process as a key component of a strategy to modernize our military forces. We are indebted to and acknowledge the contributions made by all the members of our Concept Options Group (COG): Bruno Augenstein, RAND; LT David Chelsea, USN, Naval Special Warfare; Terry Covington, RAND; Charles Duke, Los Angeles Police Department; Gene Gritton, RAND; Thomas Karr, Lawrence Livermore Laboratory; LTC Will Irwin, USA, Army Special Warfare; Douglas Kane, FBI; David Lynch, DL Sciences; Robert Moore, DST Inc.; and Alan Vick, RAND. Their experience, knowledge, and cooperation were invaluable in accomplishing this research.

We especially appreciate Bernard Schweitzer's and Alan Vick's constructive comments and suggestions, which greatly improved the clarity and accuracy of the report.

Of course, we alone are responsible for any errors of omission or commission.

ACRONYMS

AP/AM	Antipersonnel/antimateriel
COGs	Concept options groups
COTS	Commercial off the shelf
DARPA	Defense Advanced Research Projects Agency
DoD	Department of Defense
EMP	Electromagnetic pulse
EO	Electro-optical
ERPP	Effective radiated peak power
FOPEN	Foliage penetrating
GATTS/GAMS	GPS Aided Targeting/GPS Aided Munitions
GPS	Global Positioning System
HF	High frequency
IFF	Identify friend or foe
IMU	Inertial Measurement Unit
IR	Infrared
JDAM	Joint Direct Attack Munition
JROC	Joint Requirements Oversight Council
JSOW	Joint Stand-off Weapon
JSTARS	Joint Strategic Tracking and Radar System
LADAR	Laser radar
LAPD	Los Angeles Police Department
MANPAD	Man-portable air defenses
MFS	Multifrequency signatures
MLRS	Multiple Launch Rocket System
MNSs	Mission Needs Statements
MOOTW	Military operations other than war
MOTS	Military off the shelf
MTI	Moving target indicator

OSD	Office of the Secretary of Defense
PRF	Pulse repetition frequency
QRC	Quick reaction contracts
R&D	Research and development
RCS	Radar cross section
RF	Radio frequency
SAR	Synthetic aperture radar
SEAL	Sea, Air, and Land (Navy Special Forces)
SOCOM	United States Special Operations Command
SWAT	Special Weapons and Tactics
TMD	Tactical munitions dispenser
UAVs	Unmanned aerial vehicles
UWB	Ultra wide band
VHF	Very high frequency

Chapter One

INTRODUCTION

THE FORCE MODERNIZATION PROBLEM

Choosing military capabilities to be developed and purchased is always a contentious process and fraught with uncertainties. These choices are more contentious and more uncertain at times, like the present, when international developments and domestic political realities require significant changes in the character and capabilities of U.S. military forces. The only certain truth is that we cannot and should not equip the next generation of U.S. military forces by developing and buying "more of the same." The following are the reasons behind this certainty:

1. While the fundamental missions of the U.S. military change little over time, the nature of specific tasks that the military will be called on to perform is arguably changing rapidly, as are the relative emphases placed on different tasks.

2. The identity and character of potential opposing forces are also changing. Consequently, U.S. forces and the way these forces are equipped must change as well.

3. Budgetary realities will almost certainly preclude replacement of aging military systems with more modern (and often more expensive) versions on a one-for-one basis.

4. Even without budgetary constraints, we would not want to replace aging military equipment with more of the same. Emerging technologies promise dramatic new capabilities—for U.S. forces and for enemy forces—and it is unlikely that today's technological approaches will be optimal for tomorrow's military.

Equipment provided for U.S. forces and the associated operating concepts must take full advantage of newly available technology.

THE NEED FOR NEW CONCEPTS

Exploiting existing and emerging technologies to accomplish changing military tasks against diverse adversaries—and doing all of this with fewer budgetary resources—will require new approaches to a wide range of military operations. Therefore, the U.S. military must develop new *concepts* for carrying out projected missions. Of course, the need to develop new military concepts is itself not unique to the present moment; successful militaries are always adapting to changing circumstances by exploiting new technological possibilities. In times of rapid international and domestic change, however, the need to develop and to implement new concepts—and to do so quickly—becomes more acute.

This report lays out proposals for invigorating and strengthening the concept development process within the U.S. Department of Defense (DoD). It also describes efforts by RAND to test these proposals by applying them to specific military tasks.

Chapter Two of this report describes how, ideally, concept development fits into a larger process of force modernization. Unfortunately, current DoD practice falls short of the ideal, and that chapter also notes some deficiencies in the current process for identifying promising new technologies for further development. Chapter Three lays out a general framework for thinking about military operational requirements and explains how this framework was used to identify specific military tasks on which to focus RAND's concept development efforts. Chapter Four describes RAND's approach to concept development—the creation of so-called concept options groups (COGs). Chapter Five summarizes some of the ideas generated by a COG assembled by RAND to consider new approaches to military tasks associated with a hypothetical situation involving efforts to enforce a cessation of hostilities. The final chapter discusses lessons learned from the RAND experiment with a COG and offers some concluding observations. Appendix A provides a list of military activities considered by the COG. Appendixes B and C provide detailed notes on two specific topics explored by the COG.

Chapter Two

A PROCESS FOR MODERNIZING FORCES

HOW THE PROCESS SHOULD WORK[1]

The process of modernizing military forces is made up of four key steps, performed in a logical sequence by four different types of people and organizations.[2] The four steps of this process (illustrated schematically in Figure 2.1) are as follows:

Step 1: Strategic Planners Establish Demands for Military Capabilities

Strategic planners—typically within the staff of the Chairman of the Joint Chiefs of Staff and the Office of the Secretary of Defense (OSD)—are responsible for articulating the principal missions to be accomplished by U.S. military forces, consistent with the strategic goals set forth by high-level national authorities, including the presi-

---

[1]For a detailed treatment of how the framework relates to how the Department of Defense develops new operational concepts to enhance military capability, see Glenn A. Kent and David E. Thaler, *A New Concept for Streamlining Up-Front Planning,* Santa Monica, Calif.: RAND, MR-271-AF, 1993; Glenn A. Kent, William E. Simons, *A Framework for Enhancing Operational Capabilities,* Santa Monica, Calif.: RAND, R-4043-AF, 1991; and David E. Thaler, *Strategies to Tasks: A Framework for Linking Means to Ends,* Santa Monica, Calif.: RAND, MR-300-AF, 1993.

[2]For an in-depth discussion of the defense-planning process, of various techniques that can be useful in that process, and how concept development fits within a larger planning framework, see Paul K. Davis and Zalmay M. Khalilzad, *A Composite Approach to Air Force Planning,* Santa Monica, Calif.: RAND, MR-787-AF, 1996. This work uses the earlier name, Concept *Action* Groups, for what we now call Concept *Options* Groups; the activities referred to are the same.

RAND *MR912-2.1*

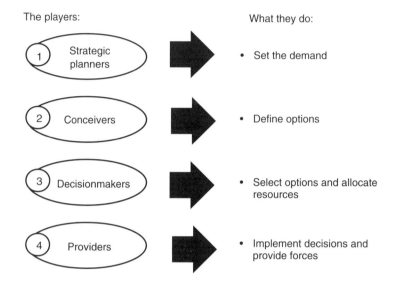

The players: What they do:

1 Strategic planners   ➡   • Set the demand

2 Conceivers   ➡   • Define options

3 Decisionmakers   ➡   • Select options and allocate resources

4 Providers   ➡   • Implement decisions and provide forces

**Figure 2.1—Force Modernization Involves Four Core Activities in a
Logical Functional Flow**

dent and his national-security advisors. Aided by operational commanders and analysts, these planners also identify the specific military tasks that must be performed if these principal missions are to be accomplished. In essence, strategic planners identify what it is that military forces have to be ready to do.

More specifically, the planners (1) set the strategic orientation of U.S. forces and describe a vision of the operational objectives to be achieved and military tasks to be accomplished by these forces; (2) establish the characteristics of military capabilities deemed most relevant to meet future needs—the operational requirements; and (3) identify deficiencies that should receive priority attention—operational deficiencies.

Strategic planners provide the terms of reference for each of the subsequent steps in the process of modernizing forces.

Step 2: Creative Conceivers Formulate Options

New concepts on which to base new capabilities do not appear automatically. They arise from the concerted efforts of creative minds that understand emerging technologies, the operational realities that military commanders face, and the overall strategic environment as laid down by the strategic planners. It is, of course, rare for this breadth of understanding to be found in a single person, and consequently the process of defining technical and operational options must be pursued by teams.

Conceivers begin the work of formulating concepts for accomplishing military tasks by thinking broadly—ideally, with no constraints other than the laws of physics. After further consideration of the technical and operational potentials of various possible approaches, some of the concepts formulated will appear to be worthy of more detailed definition in terms of particular existing or emerging technologies and specific operational tactics. Consideration of the existing state of the art and the potential for advances will result in a yet smaller number of concepts that appear sufficiently promising to justify the expense of actual proof-of-principle demonstrations. The primary product of this second step of the force-modernization process is a list of concepts that are technically feasible, operationally relevant, and ripe for further development.[3]

Step 3: Top-Level Decisionmakers Choose Among Available Options

With the aid of technical and operational evaluators, top-level decisionmakers within OSD choose which options to pursue (which concepts to implement). Costs are considered; implications for required manpower and training are delineated; the applicability of particular concepts to multiple missions is debated; judgments about the relative importance of various tasks and missions are made. This is the step where the difficult decisions about resource allocation are

---

[3]Of course, the process needs to be iterative. Often strategic planners specify needs when a technical path gives some hope of satisfying those needs. Ideas and information must percolate upward to the strategic planner as well, so that needs can make use of options enabled by new technologies.

made. Top-level decisionmakers combine the requirements laid down by the strategic planners and the options identified by the creative conceivers into choices about which missions and tasks to pursue, how to pursue them, and how many resources to devote to them.

Step 4: Force Providers Implement Decisions

The force modernization process is completed when acquisition officials act on the choices made by top-level decisionmakers, completing the development of chosen systems, procuring them, and providing them to operational commands. Force providers are also responsible for the organization, training, and tactical doctrine necessary to make the chosen new concepts and associated systems operationally functional.

Structuring the force-modernization process as described above emphasizes the *functions* that must be performed in the course of this process. Having specified these functions, we can then begin to consider the roles to be played by various *organizations* in performing these functions. Relevant organizations might include various components of the Office of the Secretary of Defense, the Joint Staff, the services, various unified commands, analytic shops, thinks tanks, and defense contractors. Thinking first about functions and then about organizations is, we believe, superior to thinking first about organizations and then defining functions to fit the needs and wishes of interested organizations.

This vision of the force modernization process also makes clear that it is the function of the force providers—who organize, man, equip, and sustain these force elements—to implement the decisions of the top-level decisionmakers. It is *not* the role of these force providers to make decisions about resource allocations.

Finally, and most relevant in the current context, this vision of the force-modernization process emphasizes the importance of formulating new concepts about how to perform required military tasks. Moreover, it underlines the fact that the objective of concept development is not to find a use for a given technology. Rather, the process of concept development begins with military requirements articulated by strategic planners and proceeds to identifying technologies that contribute to meeting these requirements. Nor is concept development concerned exclusively with hardware; formulating a concept for fulfilling a military task requires descriptions both of a technology and of how, operationally, this technology can be exploited.

CURRENT APPROACHES TO FORCE MODERNIZATION ARE LESS THAN IDEAL

Unfortunately, current practice within DoD deviates from the sequence of steps described above. In particular, current practice has the effect of reversing the second and third steps of the process. (See Figure 2.2.)

Current practice begins—as it should—with strategic planners articulating required military missions and tasks. But then, contrary to the logic of the force-modernization process, the military services (which are most naturally cast in the role of force providers, not creative conceivers) respond to these articulated needs with so-called Mission Needs Statements (MNSs), which often describe particular systems that might be suitable for meeting specific needs.

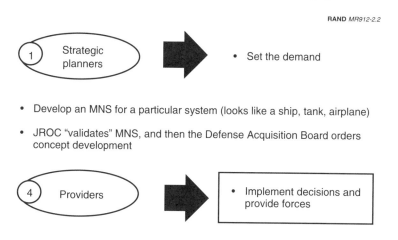

RAND *MR912-2.2*

Figure 2.2—Current Practice Reverses Concept Development and Choice of a System

Only after an MNS is validated by the Joint Requirements Oversight Council (JROC) does the Defense Acquisition Board order a formal concept-development effort. In effect, an MNS documents a choice of a particular approach to a mission and thus convolutes the process.

This reversal of steps has the effect of narrowing the range of potential concepts that receive serious consideration. In developing an MNS, a service inevitably draws on the expertise available to it, often without the benefit of expertise available in other services, in national laboratories, from contractors with which that service does not work routinely, and so on. The result is that important choices among various options are made (by default, if not otherwise) *before* a broad range of options has been formulated and considered. MNSs drafted by individual services will typically reflect the strengths of a particular service; from the outset the proposed system is likely to look like a ship, a tank, a missile, an airplane, or whatever, depending principally on which service or which part of a service is developing the MNS. By the time formal work on concept development begins, potentially interesting avenues—particularly avenues that exploit joint operations—may have been closed. The force-modernization process risks getting locked into past concepts regarding which service or which kinds of platforms will perform specific missions and tasks. Further, true competition among a broad range of alternative approaches can be precluded.

Chapter Three

THINKING ABOUT MILITARY MISSIONS

The fundamental principle underlying RAND's approach to concept development is that meaningful concept development is possible only in the context of a clearly articulated need for a type of operational capability. Without a clear statement of what tasks the military needs to accomplish, concept development can too easily become a search for a mission to be performed by an already chosen technology rather than a search for alternative technical and operational approaches to performing an already specified mission. If strategic planners have not provided the creative conceivers with clear guidance on important operational objectives and tasks, the conceivers will be tempted to fill the vacuum, concentrating on *what* rather than *how*.

A necessary antecedent to a concept development, then, is some thinking about the tasks for which we are trying to devise new concepts.

MILITARY MISSIONS

Defense planning begins with basic national-level objectives. These are found in such documents as the U.S. Constitution and are enduring and constant regardless of the geopolitical environment. Consistent with these enduring objectives, the president and his advisors set forth broad national security objectives—in the congressionally mandated *National Security Strategy of the United States* and elsewhere—toward which U.S. national power is applied. These national security objectives, which can change with international circumstances, are formulated and defined in light of U.S. interests, threats to these interests, and opportunities for advancing them.[1]

A set of national military objectives or missions is derived from these broad national-security objectives by the Chairman of the Joint Chiefs of Staff, within guidelines set by the Secretary of Defense. Broadly stated, the following are the current U.S. military missions:

- Deter and defeat attacks on the United States.

- Deter and defeat aggression against U.S. allies, friends, and global interests.

- Protect the lives of U.S. citizens in foreign locations.

- Counter regional threats involving weapons of mass destruction.

- Underwrite and foster regional stability.

- Deter and counter state-sponsored and other terrorism.

- Provide humanitarian and disaster relief to needy peoples.

Each of these missions breaks down into a number of still more specific operational objectives. For example, the following are the operational objectives derived from the military mission "underwrite and foster regional stability":

- Prevent the coercion of allied and friendly governments.

- Promote and maintain desirable regional balances of power.

- Protect threatened indigenous populations.

- Enforce the cessation of hostilities.

- Bolster democracy.

These operational objectives break down yet further into specific military tasks. For example, to "enforce the cessation of hostilities," the U.S. military has to be ready to accomplish the following tasks, among others:

---

[1]Glenn A. Kent, *A Framework for Defense Planning*, Santa Monica, Calif.: RAND, R-3721-AF/OSD, 1989.

- Stop or prevent artillery, mortars, and sniper attacks against designated targets or areas.

- Enforce no-fly zones.

- Resupply friendly forces and civilians.

- Neutralize enemy radars.

- Identify and disarm combatants.

- Locate and destroy weapons caches.

- Clear and avoid mines.

- Maintain persistent surveillance of selected areas.

It is at the level of such specific military tasks that we believe concept-development initiatives will prove most fruitful. Only when we have articulated quite specifically the nature of military tasks can we think concretely about alternative approaches to accomplishing those tasks.

CHOOSING TOPICS FOR A RAND CONCEPT-DEVELOPMENT INITIATIVE

In selecting concrete military tasks as topics for a test of a proposed approach to concept development, RAND sought to identify tasks that are militarily important, technologically and operationally challenging, and that might be accomplished through various means. More specifically, we sought to identify military tasks that have the following characteristics:

- The tasks pose significant challenges to current military capabilities and/or operational doctrines—tasks that are likely to be important but that are not satisfactorily performed today.

- A variety of modalities (e.g., airborne platforms, ground-based systems, local or remote sensors, etc.) could plausibly contribute to accomplishing the tasks.

- The tasks are not "owned" by a single service or by particular branches within services; minds are still open about who should perform these tasks and how.

- The tasks admit to a variety of technical approaches.

- The tasks have a "joint" flavor to them; multiple services or branches could conceivably contribute to them.

We applied these criteria to the specific military tasks (listed above) required to achieve the broader operational objective of enforcing a cessation of hostilities in some unstable region. (The military challenges facing NATO forces in the former Yugoslavia provide the most immediate and obvious motivation for this choice, but most observers expect U.S. forces to be called upon repeatedly in the future to perform similar missions.) Figure 3.1 summarizes how well the various military tasks met our criteria.

Of the nine military tasks considered, four met all of our criteria. From among these four, we chose two as likely to offer particularly rich sets of alternatives for discussion in the course of a concept-development exercise:[2]

- **Stop/prevent artillery, mortar, and sniper attacks against designated targets or areas.** The technical and operational challenges associated with this task include identifying the source of proscribed fires, forcing their immediate cessation, and destroying the capability for these fires to resume at a later time. Ideally, U.S. forces would prevent such fires by identifying and neutralizing threats and by preventing hostile actors from getting into position to fire. All of this, of course, must be accomplished while very strictly limiting the risk of collateral injury to noncombatant personnel and property.

- **Maintain persistent surveillance of selected areas.** The objectives of this surveillance can be manifold: recognizing movements of hostile forces, detecting the presence or proscribed movements of military equipment or supplies, identifying caches of proscribed materiel by patterns of traffic to and from such caches, verifying the withdrawal of potentially hostile forces from specified areas. The challenges in all of these cases lie principally

---

[2]These two tasks are relevant to two of the Joint Warfighting Capability Objectives laid out by the Joint Chiefs of Staff in *Joint Vision 2010: Precision Force and Military Operations in Urban Terrain,* undated (available on the Web: http://www.dtic.mil/ doctrine/jv2010/jvpub.htm).

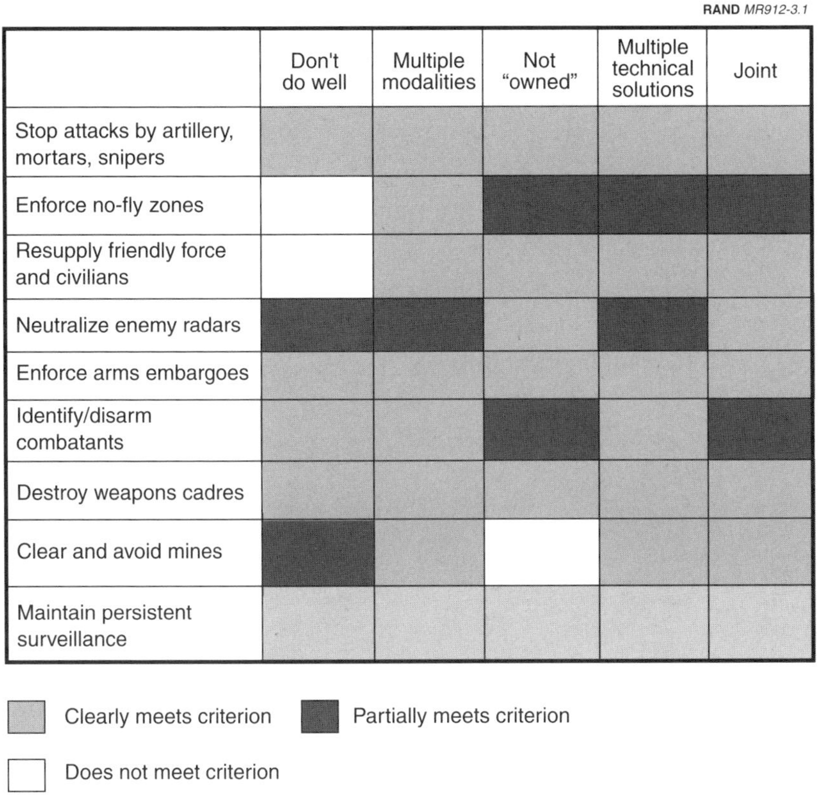

RAND *MR912-3.1*

	Don't do well	Multiple modalities	Not "owned"	Multiple technical solutions	Joint
Stop attacks by artillery, mortars, snipers	▨	▨	▨	▨	▨
Enforce no-fly zones	☐	▨	■	■	■
Resupply friendly force and civilians	☐	▨	▨	▨	▨
Neutralize enemy radars	■	▨		■	▨
Enforce arms embargoes	▨	▨	▨	▨	▨
Identify/disarm combatants	▨	▨	■	▨	■
Destroy weapons cadres	▨	▨	▨	▨	▨
Clear and avoid mines	■	▨	☐	▨	▨
Maintain persistent surveillance	▨	▨	▨	▨	▨

▨ Clearly meets criterion   ■ Partially meets criterion

☐ Does not meet criterion

Figure 3.1—Selecting Military Tasks to Analyze

in observing the brief presence or movements of small targets (small numbers of personnel, light vehicles, single artillery pieces, etc.) that will be very hard to distinguish against a background of mostly innocent noncombatant personnel, vehicles, buildings, etc.

Chapter Four

THE CONCEPT OPTIONS GROUP: A NEW FORUM FOR CONCEPT DEVELOPMENT

The central element of RAND's approach to concept development is the work of a concept options group. A COG is a small team (12 to 15 members) working intensively over a short period of time to formulate alternative technological and operational approaches to accomplishing specific military tasks. The principal objective of a COG is to provide top-level decisionmakers with a list of plausible options for future technology and systems development. The COG seeks to provide decisionmakers with a sort of "map" for future research and development efforts, identifying specific technologies that merit further development.

Successful concept development requires the combination of technological expertise, operational experience and savvy, and an understanding of the strategic environment and objectives that give rise to requirements for particular military capabilities. Consequently, we believe that a successful COG will include three different kinds of members:

- **Broadly knowledgeable technologists drawn from a variety of scientific and engineering backgrounds.** Because the task of the COG is to consider a wide range of technological options, specialists in narrow technological fields will be of limited value to a COG. Ideal participants will understand current capabilities, plausible future developments in broad technological domains, and potential countermeasures.

- **Experienced military operators.** Even the best thinking about technological possibilities will be of limited value to the military unless, from the very beginning, it is imbued with a sense of what is operationally practical and necessary. If COGs are to succeed, they must provide a forum for technologists and experienced military officers—representing all of the service branches (including intelligence) that might plausibly contribute to a specified mission—to combine their expertise.

- **A few senior analysts and planners.** These analysts and planners could come from national-security research institutions (such as RAND) or from DoD. Their role in the deliberations of a COG is to help refine definitions of the missions being addressed, provide a strategic context for the military tasks being considered, keep discussions focused on the strategic and tactical needs of DoD, and subsequently present the ideas of the COG to relevant acquisition and technology authorities.

THE RAND COG EXERCISE

In February and March 1996, RAND convened a COG to consider the two military tasks described in the previous chapter: (1) stop/prevent artillery, mortar, and sniper attacks against designated targets or areas; and (2) maintain persistent surveillance of selected areas. Our primary aim, of course, was to develop some useful new concepts for how to approach these challenging and increasingly relevant military tasks. Important secondary objectives were to gain practical experience with the operation of a COG and to derive lessons for how to make such groups more effective in the future.

To discourage a narrow focus—either technologically or militarily—we chose to ask this first COG to focus on two distinct but related military tasks. In light of the two broad topics under consideration, we needed to recruit members for the COG with sufficient breadth of knowledge to contribute to discussions on both topics.

Our initial COG met for three two-day sessions, with two-week intervals between sessions. This COG consisted of 14 members, with a distribution as shown in Figure 4.1. Seven were either RAND staff or consultants to RAND. The military operators were from the Navy

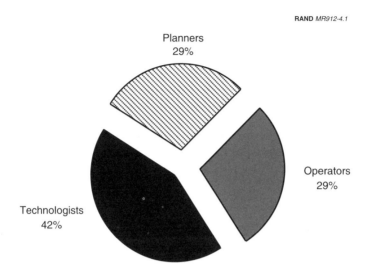

RAND *MR912-4.1*

Figure 4.1—Distribution of COG Members

Special Forces (SEAL) and Army Special Forces. For the military tasks we were considering, we felt it particularly important to include operators experienced in urban environments; so in addition to representatives from the military services, representatives from the FBI and the Los Angeles Police Department (LAPD) were invited, attended, and made important contributions.[1]

ORGANIZING COG DISCUSSIONS

We sought to structure COG discussions so as to consider five distinct questions. Inevitably, these questions sometimes overlap, but we found the discipline involved in being explicit about these questions to be helpful. Further, we believe that providing clear answers to these questions will make eventual recommendations more valuable to top-level decisionmakers. By focusing on these questions, the COG traces through the characteristics and potentialities of

---

[1]Both the FBI and LAPD representatives had extensive experience with Special Weapons and Tactics (SWAT) teams.

specific technologies, the operational context that must be created if the full potential of these technologies is to be realized, and the needs for complementary capabilities and forces.

Question 1: What, Precisely, Will U.S. Forces Have to Do to Accomplish the Military Task Being Discussed?

We began our discussions by seeking to decompose the topic military tasks yet further. For example, if U.S. forces will be required to maintain persistent surveillance of specified areas, what—in detail— will they be required to detect or to monitor? Heavy equipment? Personnel? Stationary vehicles? Moving vehicles? In what sorts of environments will these objects have to be observed or monitored? In wooded areas? In open fields? Along roads? In built-up urban areas? We did not seek to produce an exhaustive list of these very detailed military activities. (Doing so, of course, would have been impossible.) Rather, we sought to identify a handful of activities that members of the group thought to be particularly challenging given current U.S. military capabilities and to describe these activities with sufficient precision to allow detailed and concrete thinking about combinations of technology and tactics that might allow these activities to be performed better. We found this very precise definition of required military activities to be essential. Indeed, as our discussions of the technical and operational options became more specific, we returned repeatedly to the task of defining yet more precisely just what it is that U.S. forces will have to do in particular situations.

Question 2: What Are the Characteristic Signatures of Relevant Targets?

In discussing this question, we sought to develop a clear understanding of the various signatures (visual, radar, acoustic, magnetic, etc.) of objects or phenomena to be observed and monitored, the vulnerabilities of targets that might have to be attacked, and the operational steps that hostile forces would have to take (and that might be observed or thwarted by U.S. forces) in preparing for or pursuing threatening or prohibited activities.

Question 3: What Technologies Could Be Used to Exploit Signatures, Vulnerabilities, or Necessary Activities by Hostile Forces?

Having articulated in detail what U.S. forces might be trying to achieve and the characteristics of enemy forces or operations, we sought to identify various technologies that could provide a basis for U.S. sensors or weapon systems designed to counter hostile actions. In our discussions of this question, we sought to impose no restraints other than those inherent in the laws of physics. We sought to identify a wide range of ways to exploit enemy vulnerabilities.

Question 4: How Could the Technologies Identified in Discussions of the Last Question Be Exploited Operationally?

In essence, these discussions sought to sketch out an operational concept or a system concept based on promising technologies. If, for example, a particular kind of heavy vehicle can be detected with a particular kind of radar, what are plausible platforms for that radar? High-flying manned aircraft? Manned aircraft that must risk enemy fire? Unmanned Aerial Vehicles (UAVs)?

Question 5: How Can We Shape the Environment to Make U.S. Technological and Operational Options More Effective?

In pursuing this last question, we sought to broaden our emerging technological and operational concepts to include tactical and strategic considerations. Will it be easier, for example, to detect hostile heavy equipment in an area if the movement of nonhostile equipment is restricted in some way? How might this be accomplished without unacceptable limitations on peaceful activities of a population trying to return to a normal life?

PROCESS

Our considerations of these questions took on an iterative character. That is, we cycled through these questions repeatedly as we considered how better to accomplish a particular military task: We found ourselves constantly returning to earlier questions to specify more clearly, for example, just what U.S. forces might be trying to accomplish, just what characteristics of hostile equipment or forces would create signatures or vulnerabilities that could be exploited, what the capabilities of certain kinds of sensors or weapons might be, and so on. We found this iteration and continual refinement of our thinking to be essential to the process. Indeed, some technological and operational concepts that seemed plausible early in the process were discarded in the course of efforts in later rounds of discussion to define tasks, hostile forces, and friendly capabilities ever more precisely.

This process of refinement also led to our concentrating on fewer and fewer concepts. We attempted no systematic weeding out of concepts, relying instead on the sense of the group for which concepts remained interesting and promising.[2] Almost certainly, some potentially interesting and useful concepts were dropped from consideration. We do not view this as particularly troubling, however. No process can offer any realistic hope for identifying *all* potentially useful concepts related to a particular military task. A realistic goal is to identify *some* useful concepts. Allowing smart, creative people to follow their hunches is arguably the best practical method for formulating such concepts.

At no point in the COG discussions did we require or try to force a consensus. Some members of the group were skeptical about particular aspects of proposed concepts. If, however, *some* of the technologists in the group believed a particular technical approach to be feasible, and *some* of the experienced military operators found the associated concept of operations to be workable and relevant, and *some* of the strategic planners saw the package of technology and operational concepts fitting into the broader mission of U.S. forces, we deemed a concept good enough to merit further discussion and eventually recommendation to top-level decisionmakers.

---

[2]Most of this weeding out seemed to happen during the intervals between meetings of the COG. Members, it appeared, went away thinking about the various concepts that had been discussed. When they returned for the next meeting of the COG, their thoughts had focused on the subset of concepts that seemed to them to offer the most promising opportunities. We believe that allowing intervals (two weeks, in the case of the RAND COG) between meetings for individual reflection is a key part of the COG process.

Although moderators[3] attempted to guide COG discussions in a very general way, the entire process was quite informal—and productively so. As might be expected with a group of creative experts, conversations sometime drifted from topic to topic. Occasionally, moderators had to bring the group's attention back to unfinished business, but in general the instincts of the group for which topics seemed most interesting served as a perfectly adequate guide for discussion.

_____

[3]We rotated moderators for COG discussions, partly because moderating such a group is demanding and tiring, but especially because we did not want the group's considerations to be shaped by the personality or the preferences of a single moderator.

Chapter Five

SOME PROMISING CONCEPTS

The discussions of the COG convened at RAND in February and March 1996 were wide ranging, and dozens of ideas were raised for accomplishing military tasks related to maintaining persistent surveillance of specified areas and countering artillery, mortar, or sniper fire. Of these many ideas, perhaps a dozen were discussed in some detail. As discussions proceeded, four of these ideas seemed to show particular promise, and the COG sought to flesh out these ideas sufficiently to lay the groundwork for new concepts of how to accomplish specific military tasks. It is important to point out that many of these ideas are *not* new. Indeed, some were tried during the Vietnam War. However, today with the advances in sensor and computer technology, hardware miniaturization, and data processing techniques, they are more practical. In this chapter, we summarize these four concepts and characterize the current state of development of relevant technologies.[1]

---

[1]For our initial COG effort, we focused on areas in which technology choices are bolstered by prototype tests, early development efforts, and the like, to emphasize concepts that could be fielded in a relatively few years. In a more general application of the COG process, a broader technological spectrum might be explored.

UAV OPERATIONS TO MAINTAIN PERSISTENT SURVEILLANCE[2]

Recent experience in Bosnia has dramatized the need to monitor large areas for the presence of artillery or other kinds of heavy equipment controlled by actual or potential combatants, to verify the withdrawal of such equipment as part of a ceasefire agreement, and on occasion to identify and to locate such equipment with sufficient precision to allow attack by U.S. or allied forces.[3] These can be extremely demanding tasks.

- Because such equipment may be detectable only fleetingly—as it moves from one place of concealment to another, for example— this surveillance will have to be continuous.

- Because tensions may persist for months or even years, surveillance will have to be enduring.

- Because hostile forces can find cover in dense foliage, emerging only to fire a few rounds before retreating back under cover or changing position, U.S. forces will need surveillance technologies capable of penetrating foliage and tracking equipment if it attempts to move.

- Because proscribed or potentially hostile equipment may operate in close proximity to legitimate or nonthreatening civilian populations and activities, identification of potential targets must be very certain and the location of these targets established very accurately.

- Because domestic support for operations aimed at enforcing an end to hostilities may be tenuous, this surveillance must be managed in ways that limit the exposure of U.S. personnel to hostile action.

---

[2]UAVs may also need to operate in compliance with international air-traffic-control regulations and carry automatic collision-avoidance equipment such as that found on commercial airliners.

[3]A capability to perform these monitoring, identification, and location tasks will contribute to both of the military missions that were considered by the COG: maintaining persistent surveillance of contested areas and countering artillery and mortar fire.

Among the more promising concepts generated by COG discussions was a proposal to accomplish this sort of surveillance through combined operations of two kinds of UAVs, each carrying a suite of advanced sensors.

A Large, High-Flying UAV

The first component of this two-part system would be a large UAV orbiting high (perhaps 50,000 feet) above the surveillance area, out of range of man-portable air defenses (MANPADs). The role of this high-flying UAV would be to maintain continuous, all-weather surveillance over a wide area. This UAV would carry two radars:

- **A multimode microwave or millimeter-wave radar.** Operating in synthetic-aperture-radar (SAR) mode, this radar could locate heavy equipment in the open. Operating in a high-Doppler-resolution (or moving-target-indicator—[MTI]) mode, it could detect moving vehicles and the recoil motion of artillery tubes. In MTI mode, it could also track projectiles. Such radars are already operational; indeed, what is being proposed here is a smaller, less powerful version of the Joint Strategic Tracking and Radar System (JSTARS) radar. This microwave or millimeter-wave radar would be able to see through weather, but it could not penetrate foliage. In some potentially important scenarios, of course, heavy equipment may be concealed in wooded areas. Consequently, this high-flying UAV would also be equipped with the following radar.

- **An ultra-wide-band (UWB) radar for penetrating foliage.** Several experimental systems operating in the high-frequency (HF) and very-high-frequency (VHF) portions of the spectrum are currently under development and have demonstrated a capability to penetrate foliage and (to shallow depths) soil. Such radars could be used to detect structures, vehicles, and heavy equipment in forested areas as well as shallowly buried structures, such as bunkers. Although foliage is transparent to these radars, tree trunks are not.[4] Consequently, small targets are likely to be

---

[4]Hans Hellsten, *CARABAS—An UWB Low Frequency SAR,* Military Technology, May 1994, pp. 63–67.

lost in the clutter of a forest. Larger objects, however—such as trucks and artillery pieces—will produce returns that can be distinguished from the background clutter, especially if the radar is able to look down at the target at a steep angle, rather than at a shallow angle through many tree trunks.[5,6] Foliage-penetrating radars that can operate in an MTI mode have not been developed. Some technologists believe that such radars are feasible, however, and such radars could certainly prove valuable in monitoring operations.

In addition, this high flying UAV will be equipped with a differential Global Positioning System (GPS) receiver to help in establishing positions to within a few meters and a laser designator to guide precision weapons launched from other platforms against targets that can be identified sufficiently clearly to allow attack. (This kind of identification may be possible as a result, for example, of using Doppler processing to track an artillery round back to its source or to detect the recoil of an artillery tube.[7])

A large, high-flying UAV equipped in this way could provide a persistent, survivable, all-weather capability to monitor large areas and to detect the presence of heavy equipment.[8] It would also provide a capability to "pin down" particular targets—using multimode radar in the SAR mode or foliage-penetrating radar to detect a piece of equipment and then switching to MTI mode to follow that piece of equipment if it moves.

---

[5]For an excellent discussion on sensors, foliage-penetrating and other airborne radar systems, see Alan Vick, John Bordeaux, David T. Orletsky, and David A. Shlapak, *Enhancing Airpower's Contribution Against Light Infantry Targets*, Santa Monica, Calif.: RAND, MR-697-AF, 1996.

[6]Observing from a steep angle minimizes clutter caused by tree trunks, while retaining the strong dihedral returns caused by radar signals hitting the sides of trucks and buildings and surrounding terrain.

[7]This is a low probability event, given the duty cycle of the recoil and the dwell time of the MTI radar.

[8]A development challenge yet to be overcome is fitting antennas for both a multimode radar and an ultra-wide-band radar on the same reasonably sized UAV. Conceptually, both radars could utilize a single antenna, but multifunction antennas of this sort are not available today. Larger UAVs—such as Global Hawk—might be able to carry separate antennas.

In most cases, though, this high-flyer will not provide a capability for precise identification of targets, for distinguishing between pro-scribed or potentially threatening targets and vehicles or equipment legitimately or innocently in the surveillance area. Neither will it be able to track vehicles that move through wooded areas or through areas that are shielded by terrain features.

For many purposes, the high-flying UAV will detect that *something* is in the surveillance area. When a closer look is required, the high-flying UAV will cue a response from another UAV.

A Small, Low-Flying UAV

The complement to the high-flying UAV will be a smaller, stealthier[9] UAV orbiting nearby the surveillance area, over secure territory or at an altitude that keeps it safe from hostile action. When a target is de-tected by the high-flying UAV, this smaller UAV will enter the surveillance area at a low altitude—operating under the weather if necessary—to provide higher-resolution images of the target or to shadow the target as it moves through terrain that makes high-alti-tude tracking difficult. This low-flying UAV could be equipped with some or all of the following sensors:[10]

- **Low-light television** to provide visual images for target recogni-tion. Low-light television systems are already operational, hav-ing been used on AC-130s and other platforms in a number of conflicts, including recently in Bosnia.

- **An imaging infrared (IR) sensor** to allow target identification and tracking at night, in good weather, or in bad weather when close to the target. Imaging infrared sensors of at least two dif-ferent sorts are currently operational.

- **Laser radar** (LADAR) for three-dimensional imaging. LADARs small enough to be used for target recognition on terminally guided weapons have been flown in prototype versions but are not yet operational. These prototype LADARs provide sufficient

---

[9]Stealth will be hard to achieve for the low-flyer because it will be in visual range from the ground and also susceptible to IR weapons.

[10]For more detailed discussions of these imaging technologies, see Vick et al., 1996.

resolution for target recognition at ranges of about one kilometer, adequate for use on a low-flying UAV.

- **Hyperspectral sensors** to aid in reducing clutter and to capture distinguishing signatures at multiple wavelengths. Hyperspectral imaging sensors are currently being developed by several companies. Current versions have good spectral resolution but only coarse spatial resolution. Nonetheless, flight tests have demonstrated a capability to identify some targets not visible to single-spectrum sensors.

For these sensors to be effective, this small UAV will have to operate at altitudes below 5,000 feet. At these altitudes, of course, it will be vulnerable to ground fire, and its survival will depend on stealthy design, primary reliance on passive sensors (of the sensors it will carry, only LADAR will be active), and employment only in situations where adequate target identification or tracking is impossible from high altitudes. Like the high-flying UAV, this low-flyer will carry a GPS receiver and a laser designator to guide munitions launched from other platforms.

Operational Concepts

The basic operational concept for this two-UAV approach to surveillance was outlined above: The large, high-flying UAV will orbit at altitudes that provide safety from most ground-based threats and allow coverage of wide areas. The small, low-flying UAV will orbit on station outside the surveillance area and at a safe altitude until needed to provide a closer look at some target.

The large UAV will likely have sufficient range to be based well outside the area to be monitored—perhaps as much as 500 miles away. Some members of the COG proposed basing these large UAVs on board specially designed "UAV carriers"—converted cargo vessels that could be fitted with 500-foot flight decks.[11] The small, low-flying UAVs will have shorter ranges and will therefore be based much closer to the surveillance areas—perhaps within 100 miles.

---

[11]Because the long range of large UAVs could allow UAV carriers to remain well outside theaters of conflict, some COG members went so far as to suggest that these ships could be manned by civilian crews; DoD might contract out for UAV services.

Essential to the successful operation of this UAV system will be reliable, high-bandwidth communications channels for controlling the UAV and for transmitting imagery back to analysts and commanders. The high-flyer will most likely maintain communications both through uplinks to satellites and through downlinks to ground stations. Terrain features may prevent the low-flyer from maintaining continuous communication with ground stations, and downward-directed emissions may allow detection and attack. The primary communication links for the low-flying UAV, therefore, will be upward-directed through satellites and through the larger UAV orbiting overhead. Also essential will be a sophisticated "fusion center" where data coming from multiple sources can be brought together to allow effective control of the UAVs and action on the basis of information transmitted from the UAVs. This fusion center could be located almost anywhere in the world; once communication links are established from the UAVs to a satellite, information can be transmitted nearly anywhere. But wherever this fusion center is located, there must be a robust, real-time link from the fusion center to the field commanders who will have to act on information gathered by the UAVs. Designing the necessary fusion center and the associated links to commanders are challenging technical tasks that still lie ahead.

Shaping the Environment

Operations intended to enforce an end to hostilities will typically be carried out in environments where U.S. or allied forces have some ability to control conditions. By taking advantage of this leverage, U.S. forces may be able to shape the environment in ways that will make the proposed UAV operations more effective. The following are among the more important steps to be taken:

- Establish overflight rights, especially if the large UAV is based some distance—and possibly across national borders—from the surveillance area.

- Reduce the threat of hostile fire. Flights by other than U.S. or allied aircraft in the vicinity of the surveillance area will probably be restricted to allow the UAVs to operate safely and to transit safely to and from the surveillance area. Also, operation of ground-based radars in or near the surveillance area or along routes to and from surveillance areas will probably be restricted to prevent these radars from being used to track UAVs and to aim antiaircraft batteries.[12]

- Restrict movement within the surveillance area. Even the most effective UAV systems will be able to detect and to track only a limited number of targets within a surveillance area, and absolute discrimination among different kinds of heavy equipment and vehicles will never be possible. Consequently, the effectiveness of the proposed UAV operations will be enhanced if the number and kind of vehicles in the surveillance area are restricted. Particular types of civilian vehicles that might be confused with proscribed or potentially dangerous equipment may be prohibited. Nonthreatening vehicles that must transit the surveillance area for legitimate purposes might be required to pass through certain checkpoints and be issued some kind of coded identify-friend-or-foe (IFF) transponder to allow them to be distinguished from other vehicles. The operating presumption would be that an unmarked vehicle is a suspicious vehicle.

- Establish "pseudollites" to enhance GPS precision. The precision of GPS-derived coordinates can be enhanced if GPS receivers can use local terrestrial sources of GPS signals in addition to the satellite constellation that provides the basis for the system. Terrestrial GPS-signal sources established at secure and carefully surveyed locations near the surveillance area act, in effect, as artificial GPS satellites—sometimes referred to as "pseudollites."

GROUND-BASED, MULTIMODE SENSOR ARRAYS

A second approach to maintaining persistent surveillance relies on arrays of small multisensor packages. The basic concept is to seed an area of interest—necessarily a limited area like a town center, a road,

---

[12]It may be necessary to shut down civilian flight-control radars in selected areas because it can be hard to distinguish these radars from antiaircraft radars. Legitimate flight-control radars might be monitored by allied personnel to prevent their misuse, but it is possible that the only practical approach will be a total ban on certain types of ground-based radars. U.S. and allied forces may need to announce an "if it emits, we attack it" policy.

a natural chokepoint, or an area that is shielded by terrain, buildings, or foliage from other kinds of monitoring—with small, inexpensive, and unobtrusive sensors capable of detecting a variety of signatures (acoustic, thermal, seismic, magnetic, and optical). An unattended sensor relying on a single phenomenology may not be able to discriminate targets of interest against a background of innocent activity and traffic. Sensor packages, however, capable of detecting multiple signatures simultaneously may provide greater discrimination. Simultaneous detection of a particular combination of seismic, magnetic, and acoustic signatures could, for example, identify an armored vehicle or a large truck passing by.

Multiple modes of operation may also allow a sensor to become active only when an event of interest is occurring, thus conserving power and reducing information transmission requirements. For example, seismic, acoustic, or magnetic detectors incorporated into a sensor package might operate continually, cueing imaging sensors (low-light TV or imaging infrared) that require much more power and generate larger amounts of information only when a large, noisy, or metallic object passes by.

COG discussions did not focus on a single kind of multisensor package. Instead, these discussions noted a wide variety of possible approaches to combining, packaging, deploying, powering, and deriving information from sensors. The following discussion captures some of this diversity.

Sensor Packages

Unattended sensors will be valuable in situations that do not allow the continuous presence of U.S. or allied personnel—perhaps because this presence is dangerous or because available manpower is insufficient to maintain continuous presence in all areas of interest. But in these circumstances, sensors will be vulnerable to damage or removal by adversaries who would prefer certain actions to go undetected. To be effective, then, sensor packages must be small, unobtrusive, numerous, and easily (re)deployed.

Making sensors small will require miniaturization of sensor components, particularly if a single package is to contain multiple sensors.[13]

Small size will help to make sensors unobtrusive, as can placement of sensors in inconspicuous locations—e.g., high on building walls, concealed in trees, hanging from power lines. Vulnerability to detection can also be reduced if the sensor package is designed to resemble objects that are frequently found in a particular location—vent covers, rocks, bottles, etc. COG participants imagined the creation of wide variety of designs for sensor packages appropriate to specific environments where U.S. forces might be required to operate. A rapid manufacturing capability would allow packages appropriate to a particular environment to be produced quickly when it becomes clear that troops are likely to be deployed.

Sensors will also be more survivable if they do not continuously emit signals. Presumably, it will not take adversaries long to discover how sensors communicate, and a sensor's emissions could be used to locate it. Thus, an ideal sensor might be one that emits only sporadically—perhaps only when it has something specific to report or when queried by friendly forces.

COG participants described a clever technique already being used by some public utility companies for remote reading of utility meters that do not have self-contained power supplies: The meter stores information until it is prompted to respond by the receipt of a radio signal from a passing truck. At that point, the meter emits a weak but

---

[13]Our COG also discussed insect-size flying and crawling systems, capable of a wide variety of battlefield sensor missions, that could be seeded in large numbers over a target. For more details, see Richard O. Hundley and Eugene C. Gritton, *Future Technology-Driven Revolutions in Military Operations: Results of a Workshop*, Santa Monica, Calif.: RAND, DB-110-ARPA, 1994; and Keith W. Brendley and Randall Steeb, *Military Applications of Microelectromechanical Systems*, Santa Monica, Calif.: RAND, MR-175-OSD/AF/A, 1993, especially pp. 21–30. These early RAND initiatives are now generally credited with starting the current U.S. program on Micro Flying Devices; see, e.g., the *New York Times* article by W.E. Leary, "Tiny Spies that Fly May Transform War and Rescue," November 18, 1997, and the article in the *Lincoln Laboratory Journal*, Volume 9, Number 2, 1996, "Micro Air Vehicles for Optical Surveillance." This current U.S. program involves industry and academia, and is progressing toward new developments in very small vehicles, propulsion, power, and sensors. A major goal of this program is to provide vehicles with wing spans of about three inches; this reflects an intermediate step in the RAND concept leading to vehicles an order of magnitude smaller, and resulting in truly unobtrusive sensing capabilities.

detectable radio response by "retro-reflection"—the power for the response is derived from the power of the incident querying signal. If friendly forces have occasional access to areas seeded with sensors (during daylight hours, say, or during routine patrols) such techniques might allow them to receive reports of proscribed or suspicious movements of equipment that have taken place since their last presence. Because friendly forces can cover a wide area with querying signals, a casual hostile observer will have difficulty determining the location of specific sensors. The querying signal should be coded so that it cannot be replicated and used by adversaries to locate the sensors.

The simple expedient of designing sensors that will emit signals only in a narrow beam aimed upward may provide reasonably secure communication in areas where U.S. and allied forces enjoy exclusive access to aircraft.

Over time, of course, the only defense against discovery and disabling of sensors will lie in proliferation and replacement. An adversary may find and disable many sensors; but as long as some remain, monitoring is still possible. Proliferation will require that sensors be quite inexpensive, so that many can be produced.

For information from sensors to be valuable, the location of sensors must be known with some precision. The location of hand-planted sensors or precision-delivered sensors (see below) can be recorded at the time of deployment. Air-dropped or scattered sensors will have to carry some mechanism for allowing their final location to be determined. A miniaturized GPS receiver—perhaps with its accuracy enhanced by reference to pseudollites—could allow a sensor's location to be included as a part of any message it transmits. Alternatively, emissions broadcast in response to a querying signal could be used to locate a sensor by triangulation if the emission can be detected by multiple collectors; this approach increases the risk of detection by adversaries, however.

Almost all sensors will require some source of stored energy, and improvements in battery technology will be key to designing highly capable and enduring sensors. (Even sensors that rely on retro-reflection for reporting may require small amounts of power to record and to store information.) In some cases, it may be possible to deploy solar-powered sensors.

Deploying Sensors

U.S. forces charged with enforcing an end to hostilities will not always have the luxury of being able to prepare a field of operations unobserved by potential adversaries. These forces must presume that most of their actions will be observed. There is therefore likely to be considerable value in devising clandestine or at least unobtrusive methods for deploying sensors. The COG noted a number of possible approaches:

- **Disguised emplacement.** Small numbers of sensors might be emplaced by U.S. forces or local allies who appear to be doing something else: delivering goods; maintaining public property; repairing telephone lines, power lines, or damaged buildings; etc.

- **Scattering.** In rural areas, some kinds of sensors could be scattered from aircraft. Seismic, acoustic, or magnetic sensors might be attached to earth-penetrating spikes. Acoustic, magnetic, and imaging sensors might be suspended from trees or power lines. Ideally, sensors could be scattered at times and under conditions when observation is unlikely—at night, for example. Even if adversaries observe the deployment, though, they can have little assurance that they will find all of the scattered sensors. In some circumstances, sensors could be scattered along roads from vehicles.

- **Precise standoff deployment.** It may also be possible to deploy sensors by firing them from airborne or ground-based platforms. This approach may be particularly effective, for example, for placing sensors in inaccessible locations—high on the wall of a building, for example. Because this kind of deployment could happen very suddenly and because the launching platform might be some distance from the final location of the sensor, adversaries might find it difficult to observe the deployment or to locate the sensor once it is in place.

Operational Concepts

An ideal use of sensors would exploit reports from multiple net-worked sensors, all reporting the same event from multiple vantage points and possibly at slightly different times in order to detect movement. Most COG participants felt that research and development (R&D) programs could (and should) be formulated to design fully networked sensors. They also felt that achieving this ideal was far in the future, however. Extensive communications are required to operate netted sensors, and the data-processing task required to make sense of multiple signals is a formidable challenge. Further, sensor networks may be quickly degraded if some of the components fail or are disabled.

The COG thought it a sufficient initial challenge to concentrate at-tention in the near term on designing multimode sensors that could provide valuable information while operating independently. The COG also recognized that in peace-enforcing situations, real-time information might not be essential; the knowledge that certain events happened or that particular types of vehicles passed by a known location at a particular time could be valuable even if that information is not collected or retrieved until hours later. Conse-quently, much of the discussion within the COG focused on sensors that could be used to discriminate among broad classes of targets and that would report only intermittently—either when particular events took place or when queried by operators.

One attractive concept that arose in the course of these discussions was the concept of a "virtual patrol." In peace-enforcement con-texts, U.S. commanders may be reluctant to expose troops to the risks associated with routine small-unit patrolling. The forces avail-able may also be inadequate to mount frequent patrols of widespread areas. In these circumstances, even a rudimentary array of sensors—perhaps deployed and maintained during infrequent physical patrols of the area—may be sufficient to allow a technician located in some base area to keep track of what has happened or is happening in a particular area by checking regularly with many different sensors. In essence, this technician can "patrol" an area without leaving headquarters.[14]

Shaping the Environment

The principal difficulty in making effective use of sensor arrays, of course, will be distinguishing militarily significant events from background clutter. The effectiveness of sensors will be enhanced if the environment can be simplified and the number of innocent but ambiguous events reduced. Restricting movements of heavy vehicles, for example, to particular routes or to particular times of day would enhance the ability of sensors to detect unauthorized movements. The simple expedient of a curfew—arguably not unreasonable in an environment where troops are deployed to enforce an end to hostilities—may go a long way toward making sensor arrays effective. Creating chokepoints by closing roads or bridges or by maintaining regular human patrols of certain areas will allow sensors to be concentrated in fewer areas of interest.

Complete closure of specified areas on a temporary basis—even if these closures cannot or should not be permanently maintained—may allow U.S. or allied forces to deploy sensors unobserved.

Establishing supplemental sources of GPS signals will allow enhanced accuracy in locating sensors. Establishing a dense and redundant network of sites for receiving communication from sensors will add to the robustness of a sensor array.

The Current State of Technology

Although most of the basic technologies necessary for multimode sensor packages are available today, almost all will require further maturation (at the very least) before operational sensor arrays will be practical.

---

[14]This, of course, is the technique employed by security guards in large building complexes, who use multiple television monitors to "patrol" remote parts of the building.

- **Sensors:** Sensors of all the sorts proposed above are operational today, but all will require further miniaturization, cost reduction, and perhaps hardening to withstand air delivery or projectile deployment.

- **Target recognition:** Proof of principle has yet to be established for the ability of multimode sensors to distinguish targets of potential military interest from background clutter.

- **Disguised packaging:** Techniques for rapid manufacturing of predesigned items exist today but would have to be adapted to producing a wide variety of sensor packages for military use. Perhaps more demanding would be the design of a sufficient assortment of site-appropriate packages.

- **Establishing location of sensors:** GPS receivers, of course, are in common use today. Some adaptation will probably be required for inclusion in sensor packages, especially if receivers are to make reference to supplementary GPS signals from pseudollites. Proof of principle will be required for techniques that will allow orientation of imaging sensors and pointing of communication emissions.

- **Power supplies:** A difficult technological challenge associated with creating effective sensor arrays lies in creating adequate power sources. Suitably small batteries and technologies for deriving power through inductance from power lines or other nearby power sources are available in useful form but could profit from further development for application in the sort of sensor packages suggested here. Retro-reflection techniques, using querying signals to power replies, still require additional adaptation for routine operational use in the kinds of applications described here.

- **Communications:** the basic technologies for transmitting information are well understood today, but all will require further maturation to reduce power requirements (perhaps through data compaction) and to minimize chances of detection by utilizing low-duty cycle emissions or using a high frequency to narrow the beam. Further maturation will also be required for techniques to store information within a sensor and then to read it out on demand.

- **Utilizing information from networks of sensors:** Modern data processing hardware is adequate to meet the task of combining and interpreting output from multiple sensors. Considerable development work remains to be done, however, on the software necessary to allow recognition of militarily relevant events and to present information relating to such events in a form that is usable by field commanders.

TECHNIQUES FOR DETECTING CONCEALED SMALL ARMS

Snipers pose a constant threat to U.S. forces charged with enforcing an end to hostilities.[15] Even if they avoid direct attacks on U.S. forces, snipers can undermine a fragile ceasefire through attacks against civilians, destroying confidence that personal safety can be guaranteed, disrupting attempts by noncombatants to return to something resembling normal life, and perpetuating cycles of violence and reprisals.

Aggressive military responses to sniping incidents after the fact can suppress sniping activity somewhat.[16] (See the discussion below of techniques for localizing sources of sniper fire.) But in most cases, snipers will retain a distinct advantage, being able to choose the time and place of attacks and to shelter behind, or to disappear quickly among, innocent noncombatant populations—especially in urban environments.

U.S. forces seeking to enforce fragile ceasefires require methods for denying snipers the initiative, methods for identifying and neutralizing a sniper *before* he or she gets in position to fire. In particular, U.S. forces could benefit greatly from a capability to scan large groups of people for concealed weapons, so as to restrict the freedom of would-be snipers to move through urban areas on their way to or from firing positions. If this scanning could be accomplished clan-

---

[15]For more information about snipers, see Major John L. Plaster, USAF (Ret.), *The Ultimate Sniper: An Advanced Training Manual for Military and Police Snipers,* Bolder, Colo.: Paladin Press, 1993.

[16]In Mogadishu, U.S. snipers were quite effective in despatching Aideed's snipers firing from open locations. See Tony Capaccio, "U.S. Snipers Enforce Peace Through Gun Barrels," *Defense Week,* January 31, 1994.

destinely—so that a would-be sniper could never be sure whether or not he or she was being scanned—so much the better.

The current generation of magnetic metal detectors are far from ideal. Individuals to be scanned must pass through narrow choke-points one at a time, and current detectors have no capability to distinguish between weapons and a host of ordinary metal objects that would be in the possession of civilians in any urban area. Identifying a concealed weapon typically requires that scanned individuals remove all innocent metallic items before passing through a metal detector. Thus, it is hard to imagine any practical method for using existing metal detectors clandestinely.

COG discussions identified two technological approaches to detecting concealed weapons that show some promise for use in future peace-enforcement operations.

Millimeter-Wave Imaging

Millimeter-wave radiation can penetrate thin, nonmetallic material such as clothing or Sheetrock walls with little attenuation. Metallic and even some plastic weapons, however, produce strong reflections. The wavelength is sufficiently short to allow imaging at ranges of a few feet—a weapon will look like a weapon. It should be possible, then, to use millimeter-wave imagers to scan individuals passing through narrow chokepoints for weapons hidden under clothing or in nonmetallic containers or bags.[17] Because weapons could be distinguished from other metallic objects by shape, it would be unnecessary to require individuals being scanned to remove other objects, which would have alerted them that they were being scanned. It may also be possible to conceal scanning devices behind thin walls. Thus, someone carrying a concealed weapon would be vulnerable to detection whenever he or she passes through a doorway or narrow gate, along a corridor, or up or down a stairway—a poten-

---

[17]For instance, assuming a wavelength of .01 feet for the radar, a cross-range resolution of about 1/4 inch could be achieved using a 1 foot antenna at a range of 2 1/2 feet. However, achieving that resolution in the range dimension would require a transmitted bandwidth of more than 20 GHz, which is a 20 percent bandwidth for a 100 GHz radar. This may present some development obstacles that need to be explored.

tially serious deterrent for a would-be sniper. Defense Advanced Research Projects Agency (DARPA) is currently funding the development of imaging systems of this sort.

Radio-Frequency Resonance Detectors[18]

It has been speculated that concealed weapons could be detected by the hollow cylindrical metallic barrel of roughly predictable dimensions of most firearms. (Firearms will fall into a limited number of categories—pistols, sniper rifles, assault rifles, etc.—each of which will be characterized by a barrel of roughly predictable dimensions.) When radiated with electromagnetic energy of an appropriate wavelength, cylinders of certain lengths will resonate and reradiate energy with a different polarity than the incident energy. A receiver designed to recognize this polarization shift will be able to detect the presence of a metallic cylinder of a particular dimension. By varying the wavelength of the incident radiation, it should be possible to detect the presence of cylinders of a variety of dimensions. Fortunately, energy of the wavelengths necessary to produce resonance in cylinders with dimensions similar to those of a gun barrel will penetrate clothing, making detection of concealed weapons practical.[19]

Bookstores, record stores, and clothing stores routinely employ technologies similar to this to prevent theft of merchandise. A small wire of known length and thickness is attached to merchandise. On leaving the store, a customer is illuminated by a low-power microwave radar of an appropriate wavelength. A wire that has not been removed by a sales clerk is detected, and an alarm is sounded.

Resonance detectors will not have resolutions sufficient for imaging, and they will probably not allow precise location of a detected cylinder. If used to scan, say, building entrances, airport waiting rooms, or public gathering places, they could conceivably detect that *someone* in the scanned area is carrying something that has the dimensions of a rifle barrel. This could cue a more careful search or the activation of heightened security procedures. Installations in fixed

---

[18]See V. Sabio, *An Efficient Method of Target Recognition Using Spectrally Matched Filters*, Proc. of the ATR Science and Technology Conference, November 1994.

[19]Appendix C provides a more technical discussion of this technology.

locations could also allow corrections for known sources of background clutter, increasing the likelihood that a new object entering the area will be detected.

In principle, resonance detectors could be used to detect cylinders of various sizes—including, for example, artillery tubes—and, theoretically at least, resonance detectors could operate at ranges of a few kilometers. Thus, an airborne version of such detectors is not obviously out of the question and might contribute to wide-area surveillance efforts, even though the problem of a false alarm could be severe.

Operational Concepts

Millimeter-wave imagers will have to be located in narrow chokepoints through which individuals pass slowly enough to allow visual recognition of suspicious or proscribed objects. This suggests use in doorways, at airline check-in counters, or other places where people will routinely queue up. These detectors may be more effective if they are used in combination with more conventional metal detectors: A conventional metal detector will alert an operator that a particular individual is carrying a metallic object; the operator will then use the millimeter-wave imager to inspect this metallic object, ideally without the subject of these observations knowing he or she has been scanned.

Resonance detectors might allow scanning of crowds entering buildings or gathered for public events. Detection of some object with dimensions like a weapon would prompt extra security precautions or more detailed searches.

Shaping the Environment

Key to effective use of either type of concealed-weapon detector is the creation of unavoidable chokepoints where individuals can be scanned. Closing some doors of buildings so that all visitors must pass through a few entrances will help. So will routing foot traffic along a few routes, ideally ones with architectural features that could conceal detection equipment. Requiring crowds to pass through lobbies or anterooms before gaining admittance to public events would allow resonance detectors a chance to work. Such arrangements may cause inconvenience for innocent residents, but checkpoints and restricted routes are not uncommon in areas where there is a need to enforce an end to hostilities. Perhaps the additional inconvenience will not be great.

Detectors will also perform better if objects that may create false positives are banned or limited in some way. Citizens in contested areas might be warned, for example, against carrying metal suitcases or boxes, lengths of pipe, or tools that might be mistaken for weapons. Again, some inconvenience will be caused, but perhaps the very announcement of such restrictions will serve to advertise the fact that detectors will be in use and therefore to deter individuals tempted to carry concealed weapons.[20]

The Current State of Technology

As noted above, millimeter-wave imaging systems are currently under development.

Operational radio frequency (RF) resonance detectors are probably some distance in the future. Although systems using the same basic phenomenology are in widespread civilian use, proof of principle has not been established for using this technology to detect weapons barrels. Other challenging technical problems include understanding which frequencies maximize detection probabilities, reducing background clutter, and determining capabilities to detect objects at various aspect angles.

Some work on resonance detectors was done and some resonance detectors were deployed (although not very successfully) during the Vietnam War. Technological advances since then—particularly in signal processing—give hope that better detectors could be designed today. Development efforts along these lines seem to have ended with the Vietnam War, however, and many of the earlier research

---

[20]It seems likely that overt detectors would have a deterrent effect but is also likely to lead combatants to search for ways of defeating or avoiding the detectors. The COG suggested a combination of overt and covert detectors would be best.

findings seem to have been lost. Some of this work will have to be done again.[21]

LOCALIZING SOURCES OF SNIPER FIRE

No technology, of course, will allow U.S. forces always to detect snipers before they get into position to pose a threat. U.S. forces—especially those seeking to enforce an end to hostilities—will need capabilities to respond to sniper attacks against themselves or against civilians. Key to mounting an effective response to a sniper attack is a capability to determine—very quickly, because an accomplished sniper will usually have prepared a route for rapid escape—where the sniper fire came from.

If the sniper knows his or her business, this will not be easy. Silencers can make acoustical detection of the muzzle report difficult. In an urban environment, street noise and echoes off structures will further complicate the task of detecting a rifle shot or determining its source if it is detected. A well-positioned sniper can also hide the muzzle flash from optical or infrared detectors that are not very close to the line of fire.[22]

Some signatures of sniper fire are nearly impossible to hide, however. COG discussions identified three potentially promising approaches to exploiting these signatures to localize sources of sniper fire.[23]

High-Frame-Rate Infrared Imaging

Although an accomplished sniper can effectively hide the acoustic, visual, and thermal signatures of the rifle's initial report, there is nothing he or she can do to hide the projectile. As a result of aerodynamic drag, a rifle bullet in flight will heat to several hundred degrees

---

[21]Appendix C provides some specific suggestions for necessary development tasks.

[22]The simple expedient of firing through a window but from well within the room will shield a muzzle flash from observers not in the line of fire.

[23]A recent JASONs workshop (January 1995) surveyed proposed techniques for localizing sources of sniper fire. For a report on this workshop, contact the JASON project office, The Mitre Corporation, McLean, Virginia.

Celsius and will emit infrared radiation along a distinctive track. An imaging infrared detector operating at very high frame rates (200 to 500 frames per second) could detect this radiation. Readings from multiple sensors could conceivably establish the track of the projectile, plotting a vector that points toward the source.

Although many routine events produce momentary, pinpoint flashes of infrared radiation (especially in an urban environment), the flight of a bullet is sufficiently distinctive that sensors capable of detecting a bullet at several points during this flight should not suffer from a high-rate of false positives. By combining an imaging infrared system with acoustic or electromagnetic-pulse detectors (see below), the incidence of false positives should be reduced even further.

A prototype infrared imaging system for detecting bullets in flight (dubbed LifeGuard/DeadEye) is currently under development at Lawrence Livermore National Laboratory. The current system is effective in plotting a two-dimensional bullet track; netted observations from two or more detectors might produce a three-dimensional plot. This kind of system is not spoofed by countermeasures and is not confused by acoustic echoes or background clutter. The system is passive, which is important; it gives off no signal that indicates its presence—no signal that could be exploited by an adversary seeking to attack it.

Acoustic Detection of a Bullet in Flight

Another undisguisable signature of a bullet in flight (at least one moving at supersonic speed) is the sonic shock wave it creates. (This is the "crack e" sound of a bullet passing nearby.) Probably detectable over shorter ranges than a bullet's thermal signature and more easily lost in echoes and background clutter, this acoustic signature might nonetheless permit plotting of a bullet's trajectory that passes close enough to two or more netted microphones. Some early development work on such systems is currently under way in the United States. French forces in Bosnia deployed an experimental version of such a system, with microphones mounted on opposite ends of an armored vehicle. Like infrared sensors, acoustic sensors have the advantage of being passive, giving no evidence of their presence or operation that can be detected by adversaries.

Detection of Electromagnetic Pulse (EMP) Associated with Rifle Fire

The most speculative approach to localizing sources of sniper fire relies on the fact that at least under some (perhaps all) circumstances, the plasma of hot gasses produced in the breech of a rifle creates a detectable electromagnetic pulse. The character of this pulse is not well understood, and it is not clear to what extent it can be distinguished from background electromagnetic emissions, over what ranges it can be detected, or whether direction to the source of the pulse can be established. Apparently, some poorly instrumented experiments were conducted in the early 1970s on EMP phenomena associated with weapons firing,[24] but COG participants were unaware of the quality of these data or what has become of them.

Perhaps the best use of EMP detectors (if they turn out to be feasible) will be in combination with infrared or acoustic detectors. By measuring the time between observation of an EMP and observation of a bullet in flight, some estimate of range to the source of the fire can be calculated.

Operational Concepts

The operational ranges of the sorts of detectors noted above are not well understood. For practical purposes, however, their use in urban environments will almost certainly prove very local. Both imaging infrared and acoustic sensors, for example, must have direct line-of-sight of at least a part of a bullet's trajectory to establish its track. Neither can "see" around corners or through buildings. Thus, such systems will be effective only if deployed near the source or the targets of sniper fire. In some cities where sniper fire is concentrated in a relatively few areas—as in "Sniper Alley" in Sarajevo—this may not prove to be a particular problem. But in circumstances that allow sniper attacks in many different parts of a city, "seeding" enough sensors to give a high likelihood of detecting most sniper fires will be impractical.

---

[24]JASON antisniper workshop, 1995.

But there may be little point in localizing sniper fires if there are no friendly forces in the immediate vicinity to respond promptly. Presumably, there is little value in knowing exactly where a sniper has fired from if he or she will be gone by the time friendly forces arrive. Thus, it may make sense to design sniper-detection systems to be mobile, so that they can move with friendly forces. If the detectors can localize a sniper's position, forces are immediately on the scene to take countermeasures. (This was the philosophy behind the mobile French system deployed in Sarajevo.) In extreme situations, sensor and weapon could be linked to automatic return fire. (The Lawrence Livermore DeadEye systems does this.)

To make these detectors mobile, some method must be found for accurately establishing the location of the sensors and the relative positions of multiple sensors if more than one is used to establish the track of a projectile. Relative position of sensors is most easily established by mounting multiple sensors at distal points of a single platform (as with the experimental French system), but this will limit the flexibility of systems and make them hard to disguise or to hide. A better solution might be to equip each sensor with a GPS receiver exploiting local terrestrial signal sources to establish positions differentially. Even if multiple sensors are mounted on a single platform, that platform will require some means of establishing its location and orientation. It may turn out to be only minimally more demanding to provide each sensor with its own location-establishing device.

Another challenge will be to instantly translate bullet trajectories into tactically meaningful terms for operational forces. At a minimum, a device might point (like a compass) toward the source of a detected fire, but this will provide a meaningful azimuth only if the pointer is located somewhere close to the path of the bullet. Indicating the correct azimuth from a point offset from the path of the bullet will require an estimate of range in addition to the orientation of the track. A valuable adjunct to a system capable of establishing the path of a projectile, therefore, would be a detailed digital map of the surrounding area showing the locations and heights of specific buildings. With such a map, it might be possible to identify the first building, say, that lies along the plotted track of a bullet. Rather than just pointing out the direction from which a bullet came, a display might be able to highlight on a more-or-less normal map a particular building as the likely source of fire.

The utility of such digital maps of areas of operations—richly detailed, easily updated, rapidly distributed to tactical units, and accessed by a variety of software applications—was repeatedly noted by COG participants in conjunction with a wide variety of technological possibilities. Various approaches to creating, updating, distributing, and using such maps might themselves constitute appropriate topics for a future COG. (Private-sector initiatives related to digital mapping may contribute to developing these capabilities.)

Shaping the Environment

The principal environmental actions necessary to enhance the effectiveness of systems designed to localize sources of sniper fire have already been noted:

- Where possible, deploy sensors in fixed locations where sniper fire is particularly likely or particularly problematic.

- Establish terrestrial sources of GPS signals to allow mobile sensors to establish locations accurately.

- Prepare detailed digital maps of areas to be monitored by these systems.

Chapter Six

OBSERVATIONS

To succeed, the force-modernization process will need to provide new ideas with technical and operational reality. In a literal sense, the aim of the force-modernization process is to empower new ideas. Obviously, to empower new ideas we must, in the first instance, formulate new ideas. The purpose of this RAND effort has been to test and to demonstrate a particular approach for formulating new ideas—in our view something of a neglected art form.

The RAND approach centers on convening a group—called a concept options group—and then causing interactions among the members. Initially, the interactions focus mainly on formulating and defining new concepts to accomplish stated tasks. Subsequently, though, we reverse the process, assessing the relative merits of the various concepts defined in accomplishing a variety of tasks.

The participants in the experimental RAND COG process were selected from diverse backgrounds—scientists, engineers, analysts, military operators, and generalists. A series of three meetings was planned and conducted, each meeting lasting two days, with two weeks between each meeting.

The actual concepts developed by the RAND COG are described in Chapter Five of this report. In this chapter, we relate observations about the process itself.

HOW THE GROUP FUNCTIONED

By design, the group addressed a particular "task to accomplish" several times—about two or three times within each of the three meetings. From previous experiences, we expected changes regarding how to accomplish the tasks—both from individuals and from the group as a whole. This dynamic held true in this particular experiment. As we proceeded, there were notable changes or refinements by most of the members in their ideas of the best concept and how effective that concept might be.

This dynamic (the lack of early closure) underlines two observations: (1) there is considerable utility in causing repeated interactions on the same matter; and (2) it is unwise to settle on the first ideas presented. Interaction among the members made the participants think differently and more perceptively, and thus often caused members to reconsider and to reshape their ideas.

Each of the different types of participants made important contributions:

- The operators were essential in continually framing the problem (task), and this was necessary in keeping the engineers, scientists, and generalists focused. Also, the operators were effective in quickly weeding out concepts that were not operationally viable.

- Without the scientists and engineers, the group would not have known about many new options—options soon to be available based on emerging technologies.

- Analysts and generalists played a key role in demanding that the participants were not finished "defining a concept" until the group could describe the concept with some degree of coherence all the way—from "finder," to "controller," to "shooter." (More about this later.)

All of the above underlines the utility of having considerable diversity among the participants.

We found it quite useful to combine two different approaches in structuring COG discussions:

- First, announce that for the next several hours we will address a specific "task to be accomplished" and entertain alternative concepts for accomplishing the stated task; and then repeat this for other tasks.

- Later, reverse the perspective. List the more promising concepts stemming from the first effort and entertain ideas about the role of that particular concept in accomplishing various tasks.

This iteration (reversal) proved to be quite useful in assessing the overall value of various concepts and should be an explicit part of the agenda for any future COGs.

We encountered few difficulties in *formulating* new concepts. There was some difficulty, however, in enforcing a discipline about identifying the more interesting concepts once formulated. To do so, one is obliged finally to be quite explicit and precise about defining the concept end-to-end: Exactly what event is being observed and with what sensor; at what range can the sensor make this observation; how does the sensor know what it is observing; how many sensors do we need; how did we get the sensors in the right place at the right time in the first place; what does a sensor report, to whom, and how; what is the response to the event being observed; did the sensor locate the target accurately enough to enable a particular response; are the time lines from observing the event to executing the response compatible with what we are trying to accomplish; how did we get the "responder" (the shooter) in the right place at the right time; what do the critical hardware components attendant to the concept weigh and about how much do they cost; and when will the various components and systems be available?

Some of these questions could be answered in some detail. However, in many cases only rough estimates were available. This suggests that we should not expect too much from the COG—a group convened for formulating or spawning new ideas; i.e., we should not necessarily expect this group to go very far in *detailing* these concepts—especially with respect to "doing the numbers."

HOW WELL DID WE DO?

How well did this particular COG accomplish its purposes?

The group did quite well in formulating and expressing new ideas and concepts. The menu of options was impressive even though certainly not exhaustive. We knew about the options discussed at the meetings. By definition, however, we were ignorant about what we did not know.

The interactions within the group went a long way in informing the various members about matters not directly associated with their backgrounds. The operators learned a lot about the functional capabilities of emerging technologies, and the scientists and engineers gained a much better insight into and respect for the tasks and challenges presented by the operators. Such a learning process is surely of some value to both—especially in the long run.

As already noted, the COG was somewhat weak with regard to "numbers." In retrospect, this should have been expected, and it suggests the need for a two-step process: (1) Use the COG for formulating and spawning and for providing very rough numbers; (2) subsequently, charge engineers to refine the numbers—certainly before recommending demonstrations of a particular concept or advocating that resources be allocated to implement the concept.

HOW THE PROCESS MIGHT BE IMPROVED

Some of the members of the group had been involved in previous efforts of a similar nature. It is not surprising that these members brought the most energy to the interactions. Accordingly, there should be increased attention toward recruiting these types of individuals for future groups. To begin to have any confidence of a reasonable grasp of the complete menu of opportunities, three or four fountains of knowledge about new technologies are required. This suggests a career field for some technologists; the more often they serve, the better they are, and the more they are in demand.

The moderators did a good job in keeping the discussions focused on the right track and moving down that track. However, the moderator was also the writer and spent precious time writing on a blackboard. It would seem that a separate person recording the discussions could greatly expedite the process. Also, using a higher-tech approach—typing on a keyboard and projecting the prose on a screen for all to see and revise—would be more efficient.

At times there were false starts with regard to how to organize the discussions of the group. Although the organizers of the effort had considered how best to organize discussions, group dynamics forced changes. This suggests that interacting with *several* people before deciding on any particular approach might be helpful in avoiding false starts.

In addition, we heard too little from the operators. In future COGs, we need to draw more insights from them, not only about experiences in actual operations, but also their candid evaluations about concepts formulated.

DETAILED MILITARY ACTIVITIES CONSIDERED BY THE RAND COG

The following activities relate to maintaining persistent surveillance of selected areas:

- Monitor a separation zone.

- Detect heavy weapons, vehicles, and personnel.

- Monitor containment zones for heavy, indirect-fire weapons and detect such weapons outside the zones.

- Monitor roads and detect roadblocks.

- Detect threatening chemicals and precursors.

- Detect armed individuals in a crowd, in a building, and passing through a checkpoint.

- Identify passive communication devices, radars, night-vision devices, range finders, and antennae.

- Report information in a timely manner.

- Identify potential blind spots.

- Detect population movements (refugees and troops).

- Monitor airspace and waterways.

- Detect equipment concentrations and movement.

- Find underground corridors of various kinds.

- Detect and identify training activities and exercises.

- Monitor potential sniper positions.
- Track hostile leadership.

Activities relating to stopping/preventing artillery, mortar, and sniper attacks against designated targets or areas follow:

- For intelligence preparation of the battlefield, identify potential firing points.
- Detect the attacker before he or she gets into position and sets up, including detection during transit of controlled areas.
- Neutralize the attacker before he or she causes harm, including detecting the setup and (in the case of a sniper) detecting and dazzling the gun sight.
- Neutralize the fired round.
- Kill the attacker after the shot (requires timely detection and location of the shot and may require low collateral damage).

COUNTERBATTERY SENSOR SYSTEM CONCEPT[1]

INTRODUCTION

U.S. forces have a need for improved sensor systems with an effective capability to detect, identify, and engage indirect-fire weapons, including artillery, multiple rockets, and mortars. (Ballistic missiles are not considered here because they are covered by the Ballistic Missile Defense Program.) These indirect-fire weapons are available and are in the forces of all but the most primitive armed forces of the world. Indirect-fire weapons have many uses, ranging from suppression and destruction of infantry antitank weapons to acts of "terrorism" and harassment against civilians of an opposing regime.

These weapons are difficult to locate and to track for long enough periods to engage because of their range and mobility. Their range places them in enemy territory and permits them to be normally located in concealed positions or under camouflage. Terrain and foliage are used for concealment. Their mobility permits them to move quickly from concealed positions to firing positions and then, after firing, to move to new firing or hiding positions ("shoot and scoot" or "shoot and hide"). Weapons may be placed in populated areas near churches, schools, hospitals, or in other areas where counterfire is restricted.

Tactical air operations are not yet very effective for locating and destroying hostile weapons, especially from safe standoff ranges or alti-

---

[1]This appendix is based largely on notes provided to the COG by COG member Robert Moore, DST Inc., Arlington, Virginia.

tudes when air defense and ground fires are present. Ground-force operations require the deployment of heavy equipment and personnel deep into the crisis or conflict area.

We have essentially no capability to locate and suppress belligerent hostile weapons without high risk of casualties inherent in overflight and insertion of people on the ground. This appendix defines a concept, possible with emerging technology, that would enable target detection and designation of hostile weapons for precision engagement by standoff weapon systems.

LIMITATIONS OF CURRENT HOSTILE WEAPON LOCATION SYSTEMS

We have few specialized systems for counterbattery operations and those that we have are not usable and effective in all situations. The AN/TPQ-36 Firefinder radar was developed to locate hostile mortar and artillery fire and to register friendly fire. It does this by tracking the projectiles and using a simple ballistic model to deduce the firing location. The accuracy depends upon the portion of the trajectory that is within the radar line of sight; it is normally good enough (on the order of 50–100 meters) for *suppressive* counterfire with artillery and the Multiple Launch Rocket System (MLRS) firing warheads with antipersonnel/antimateriel (AP/AM) submunitions dispensed over large areas.

The AN/TPQ-37 Firefinder radar is larger than the -36, with greater sensitivity, coverage, and scan capability to track longer-range artillery projectiles for counterfire targeting. The Firefinder radar's range is about 40 km, which may be extended out to 100 km to give it a capability to detect tactical munitions dispensers (TMDs) before the smaller and harder-to-detect submunitions are released.

Army doctrine requires that Firefinder radars be sited on prominent terrain to get the screening crest as low as possible. In a city or town, the radar may have to be airlifted onto the top of a building to gain adequate coverage of the surrounding area. In mountainous terrain it is difficult to obtain low and consistent screening crests. If the radar's field of vision is too high, accuracy is poor and belligerents might fire under the radar coverage. If the field of vision is too low, ground clutter may be too great for early detection and accurate tracking.

There have been and will continue to be situations wherein our military capabilities will be needed to stop or to suppress hostile weapon fires without the commitment of tactical air assets or ground force insertion. Effective suppression of hostile fires typically requires quick reaction. But historically, quick reaction has not been possible when the lives of our people are at risk. Thus, the suggested concept emphasizes UAVs for detection and targeting, along with standoff systems for engagement.

ADVERSE-WEATHER UAV BATTERY LOCATION SYSTEM

Operational experience with UAVs, notable over the last five years, demonstrates the necessity of operating in or above weather, unless the UAV is quite stealthy or low enough in cost to be considered expendable. The initial Predators deployed for Bosnian operations were equipped only with electro-optical (EO) sensors. Bad weather and low ceilings forced frequent low-altitude operations under the clouds, where UAVs were vulnerable to ground fire. Two Predators were shot down, UAV operations were suspended. Some UAVs were fitted with SARs and were redeployed.

While this Ku band SAR will allow the UAV to see through clouds from a safe altitude, it cannot penetrate foliage or camouflage. Hostile weapons are either firing, hiding, or moving to a new hiding position. In scenarios of greatest interest for the future, they will be hiding most of the time. Unless we are always in position to detect these weapons whenever and wherever they might fire, we must have a capability to find them in hiding positions, normally under or in foliage. The quest for foliage penetrating (FOPEN) radars eluded us during the Southeast Asia Conflict but is now within reach because of the advent of UWB radar technology. Several experiments are being conducted with UWB radars. Typically, they operate at UHF, have bandwidths of 300–500 MHz, and operate in the SAR mode. Range and cross-range resolutions are about one meter, which is sufficient to enable identification for some targets.

In addition to the FOPEN radar, the UAV should be equipped with a multimode microwave or millimeter-wave radar (depending upon the size of the UAV selected).   This radar would operate in the Doppler (MTI) mode as well as in the SAR mode. In the MTI mode, its functions would be to detect and to track projectiles, to detect artillery tube recoil, and to track hostile weapons scooting after shooting. In the SAR mode, it would have limited FOPEN capability to complement the UWB radar and would verify recognition. Operating in both modes this radar would enable target "pindown." In pindown, the target is monitored in the SAR spotlight mode and, when it moves, tracked in the MTI mode. If it moves into a hiding place, e.g., a building, culvert, cave, or heavy impassable foliage, its position is still known. SAR and MTI radars have been designed and demonstrated for a number of UAV programs.

These two primary sensors, along with other mission equipment, would be integrated on a UAV, such as Predator or possibly the Tier 2+, depending upon weight, volume, and endurance requirements. The former may be too small but the latter is certainly of sufficient size. A complete EO sensor package with visible and infrared imaging sensors might be desirable for use under the limited conditions when there is clear line of sight. Certainly, the payload should include a laser designator for precision delivery of laser-guided ordnance by strike aircraft, either fixed wing or helicopters. In addition, the laser would have a range-finding mode so that the range, bearing, and elevation to the target can be determined in GPS coordinates. This function requires that the UAV be equipped with a GPS that would enable the delivery of Global Positioning System/Inertial Measurement Unit (GPS/IMU) weapons, such as Joint Direct Attack Munition (JDAM) and Joint Standoff Weapon (JSOW), in the differential mode to an accuracy of as good as 2–3 meters. SAR could also be used to enable GPS weapon delivery in the differential mode by using the SAR to locate the target with respect to the UAV. The Air Force GPS Aided Targeting/GPS Aided Munitions (GATTS/GAMS) concept operates in this mode.

This UAV counterbattery concept requires on-board SAR processing and a multisensor fusion system as well as a downlink and uplink for communications and control. The processing is technically feasible but will be challenging in terms of packaging.

RESONANCE WEAPON RECOGNITION[1]

INTRODUCTION

An important objective in "military operations other than war" (MOOTW) is the detection of shooters, both before and during firing. The most common weapons used by individuals through division-sized units are tube-launched projectiles. One possible method for recognizing projectile firing weapons is resonance detection. Weapons such as artillery, mortars, tube-launched rockets, rifles, and pistols all have hollow cylindrical barrels made of metal. They usually have diameter-to-length ratios of 5:1 or greater, so the resonance is fairly narrow. The use of resonance to detect weapons is not new; it dates back at least to the Vietnam era. At that time, there were several quick reaction contracts (QRCs) that explored electromagnetic resonance phenomena for the detection of everything from tanks to bicycles to rifles under jungle canopy on the Ho Chi Minh trail. Although these techniques worked most of the time, performance with military utility was limited by the signal processing and microwave technologies then available. To the author's knowledge these methods were terminated at the end of the war. Subsequently, the government's concern for the unauthorized disclosure of classified Vietnam documents resulted in the directed destruction of massive amounts of technical data from World War II through Vietnam. Very little information seems to exist from those programs, and some of the basic measurements probably will need to be repeated if new ef-

---

[1]This appendix is based largely on notes provided to the COG by COG member David Lynch, DL Sciences, Northridge, California.

forts are undertaken. The limitations of those earlier efforts arose from inadequate false-alarm control, instabilities in analog circuitry, insufficient bandwidth and frequency stability, limited signal processing gain, poor waveform choices, and virtually nonexistent tactical digital computers and associated software. The required technologies have advanced dramatically since then, and military utility can now be achieved.

RESONANCE WEAPON DETECTION

Every young child is familiar with acoustic resonance effects through drums, whistles, blowing over bottle openings, singing into pots, striking glasses with spoons, and so on. Many adults are aware of, or are familiar with, electrical/electromagnetic resonance if they use a radio or TV antenna, magnetic resonance imaging, a laser, or if they wonder about the colors in an oil film on water. Libraries and many retailers of books, records, tapes, and compact discs use resonant wires concealed in the above items in conjunction with low-power microwave radars at their doorways to control theft. Although details of each of these phenomena are very different, they usually depend on the ratio of illuminating wavelength (λ) to some object dimension (L) and the object orientation relative to the illumination or polarization. The resonance region typically is defined as $1 < L/\lambda < 10$. Usually the resonance peaks occur near $L = n \times \lambda/2; n = 1,2,3...$, as shown in Figure C.1.

Referring to Figure C.1, a rifle barrel 1 meter long will have a peak radar cross section (RCS) of a little over 2 meter^2, at a frequency of 570 MHz. This frequency easily penetrates fabric, vegetation, windows, fog, rain, tunnel entrances and some plaster, brick, doorways, and structures. Since all weapons are not the same length and diameter, a wide range of frequencies must be used to detect a specific class of weapons, such as rifles. Such wide radar bandwidth is much easier to achieve today than it was 25 years ago and is achieved routinely in RCS instrumentation radars.

Electrical/electromagnetic resonance of cylinders depends on propagating surface currents and their reflection or reradiation at boundaries or discontinuities. Although there is some loss in a conductor such as a steel barrel, the equivalent scattering cross section is dominated by energy conservation at discontinuities. This results in

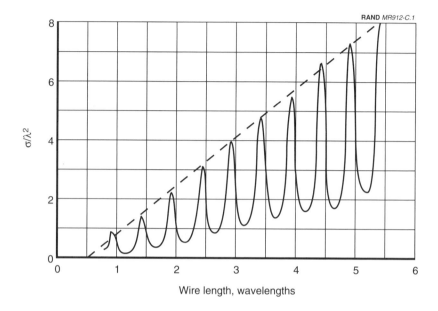

Figure C.1—Normalized Radar Cross Section of a Thin Cylinder or Wire

mode conversion so that some scattered energy is reradiated with a different polarization than that of the illumination. One powerful signature discriminant is the ratio of the like-polarization to the cross-polarization scattered from an object. In addition, multiple measurements averaged over a range of object aspect angles are usually required to discriminate objects.

Since weapons in a particular class have variations in length and diameter, illumination will be required at a range of wavelengths. The largest weapons require the lowest frequencies/longest wavelengths. For example, a 155 mm howitzer with a 10 ft barrel has a resonance region from approximately 100 MHz to 1 GHz for the barrel length and 2 to 20 GHz for the barrel internal cavity. Similarly, for a 7.62 mm rifle with a 30 inch barrel, the regions are 400 MHz to 4 GHz and 8 to 80 GHz. Observe that there is some overlap for weapons that are quite different. There are commercially available RCS measuring radars that have instantaneous operating bandwidths of 200 to 800

MHz, 2 to 4 GHz, and 8 to 12 GHz. They will provide signatures adequate to detect and discriminate a wide range of barrel weapons.

The next natural question to ask is "At what range can detection and recognition of weapons take place with safe radiation levels?" Several scenarios must be assumed to determine such an answer. One scenario is utilization of a radar in an airborne platform operating at 5,000 feet above ground level. A second scenario is placement of a radar on the roof of a four-story building surveying streets and open squares where people have congregated over a 2:1 range in distance. U.S. radiation safety limits are 1 mw/cm^2 for non-ionizing radiation. For the cases of interest, the RCS of weapons will be assumed to be $\sigma = \lambda^2$ (it will usually be larger depending on choice of λ). In the first scenario, the effective radiated peak power (ERPP) will be set to provide 1 mw/cm^2 at 5,000 feet. In the second case, the ERPP will be set so that 1 mw/cm^2 will occur at $R_{max}/2$.

The governing radar equation is:

$$R_{max} = \frac{P_{safe} \times \lambda^2 \times A_e \times G_r \times L_{prop}}{4 \times \pi \times K \times T_0 \times B_r \times NF \times SNR}$$

where:

G_r	=	29.5 dB signal processing gain (see description below)
L_{prop}	=	propagation losses, negligible for ranges achievable
P_{safe}	=	1 mw/cm^2, safe power density
A_e	=	5 m^2 aircraft, 10 m^2 ground, effective antenna area
λ	=	0.6, 0.1, 0.03 m, wavelength for the 3 bands of interest
B_r	=	500 MHz, receiver bandwidth, should be sufficient about any center frequency (selected by weapon class)
SNR	=	16 dB signal-to-noise ratio required for high detections and low false alarms
NF	=	4 dB, noise figure, typical of current state of the art
$K \times T_0$	=	204 dBw/Hz, Boltzman's Constant times Temperature

Essential to the success of mansafe weapon detection is a large signal processing gain so that small transmit powers can be used. (Safe power densities were not a consideration during the Vietnam War.) The signal processing gain arises from the assumptions about the waveform. The waveform selected is a range-gated high pulse repetition frequency (PRF) with a 100 nsec pulse, 32 rangebins, 300 kHz PRF, 30 msec minimum coherent dwell time (dependent on frequency, scan rate, velocity, ground or airborne, etc.). Range-gated high PRF is used on several modern radars, such as the radar in the F-22. The intrapulse modulation is a pure FM waveform generating 49 equispaced tones, 10.2 MHz apart, and approximately the same amplitude (a similar waveform was patented by E. Gregory et al. and has been used in radars built in the late 1970s and early 1980s). Each tone is separately filtered, range gated, and coherently processed for the full dwell time (also demonstrated in the late 1970s and early 1980s). The net integration gain for the 9,000 pulses in a dwell will be approximately 29.5 dB. The AWG-9 radar designed in the early 1970s has this much processing gain. Two orthogonal polarizations are processed on receive, and a discriminant is formed from the 98 degrees of freedom in each rangebin. Multifrequency signatures (MFS) of the type described here are used on such airborne radars as the APQ-70 and APQ-71.

The resolution cell size is large enough that target-to-clutter ratios will often be unfavorable. Therefore, clutter rejection techniques are required. In the airborne case, Doppler beam sharpening can be used to create 15-by-15-meter cells in which to detect signatures. For the ground case, a slow scan aperture and a Kalmus-type clutter filter are required. A rifle must be moving at perhaps one mile per hour for rifle-sized targets. The ground system described would provide 15-meter range resolution and about one-kilometer cross range—so it is not an imaging radar. The resolution cell size is similar to early periscope and mortar locating systems, which were adequate for detection.

Once detected, angle discriminants can be formed on detected targets to track individual weapons. The three frequency bands might be used simultaneously or sequentially, depending on required update rates. MFS techniques provide excellent recognition for many classes of targets. However, these technologies are unproved for projectile weapons. Another area that requires system analysis is the frequency used versus the achieved background contrast, i.e., the radar cross section of the target class goes up as the frequency increases, but the signature also becomes more directional.

Substituting into the radar equation using dB notation for the six cases above yields the results shown in Table C.1.

RECOMMENDATIONS

Although the set of techniques described above will not be easy to develop, there are substantial reasons to believe it is feasible. Developing such a capability will require some phenomenology experiments, signature measurements, and substantial engineering. The programs necessary to develop this capability consist of the following three generally successive steps:

- An initial feasibility study.

- A program to develop a quick reaction demonstration capability.

- A program to field an operational capability, based on a successful demonstrator capability.

Table C.1

Results of Substituting into the Radar Equation Using dB Notation

Parameter	Airborne	Airborne	Airborne	Ground	Ground	Ground
λ, m	0.6	0.1	0.03	0.6	0.1	0.03
λ^2, dB m^2	−4.44	−10	−15.23	−4.44	−10	−15.23
A_e, dB m^2	7	7	7	10	10	10
G_r, dB	29.5	29.5	29.5	29.5	29.5	29.5
P_{safe}, dB w	−30	−30	−30	−36	−36	−36
Kt_0, dB w / hz	204	204	204	204	204	204
$-B_r$, dB hz	−87	−87	−87	−87	−87	−87
$-Nf$, dB	−4	−4	−4	−4	−4	−4
$-SNR$, dB	−16	−16	−16	−16	−16	−16
-4π, dB	−11	−11	−11	−11	−11	−11
R^2, dB km^2	88.06	82.5	77.27	85.06	79.5	74.27
Range, km	25.3	13.3	7.3	17.9	9.4	3.7

NOTES: All of these ranges are operationally useful. Accounting for internal losses could modify these results.

These programs are assumed to be so phased that termination or advancement decisions can be based on findings in each of the individual steps.

Note, however, that with careful planning these steps can have some overlapping concurrency, particularly in steps 1 and 2.

Step 1. Conduct a Feasibility Study

Conduct a careful literature survey; perform phenomenology tests of five common rifles (e.g., M16, AK47, AR14, M1 carbine, etc.), three common mortars, and the barrel and breech of two artillery pieces on a high resolution RCS range; perform system analysis based on signature phenomenology; generate strawman designs for potential hardware realizations; perform analysis and simulation of strawmen; recommend a feasibility system. The span of performance may be 6–12 months.

Emphasis in the first step can be placed on maximal restoration of the past database, including, where possible, interviews with past researchers and others knowledgeable in this field; careful analysis and categorization of possible signature discriminants (e.g., polarization ratios); and attention to identification and characterization of those environmental and target characteristics that may differently affect operational utility of these resonance detectors, assessing for various radar types such a priori predictions as the following:

a. If the terrain is benign and the aspect angle of the cylinder is favorable, can one detect the object with reasonably high probability and a tolerable false alarm rate—provided the target is effectively in the open?

b. If the background is challenging (wooded or urban environment), how good a localization cue is required to keep the number of false alarms within reason?

c. Choosing the best frequency can be a compromise between enhancing the radar signature of the target on the one hand, and decreasing clutter and propagation attention (e.g., because of foliage penetration) on the other. The optimal resonance frequency may turn out not to be optimal for the overall situation.

Step 2. Develop a Feasibility Demonstrator

Develop a feasibility system based on selected configuration of step 1; the demonstrator would be based on maximum use of commercial-off-the-shelf (COTS) and military-off-the-shelf (MOTS) hardware and software (e.g., adapt a commercial RCS instrumentation radar); conduct RCS range tests to demonstrate successful target classification; conduct field tests in a battle lab environment (e.g., Fort Irwin, etc.); create and expand a signature database; recommend a deployable system. The span of performance including field tests should be 18–24 months. At the end of this phase, a user-operational evaluation system would exist for deployment.

A substantial emphasis on step 2 would be the selection of "breadboard" componentry adequate to test major conclusions of step 1, including those environmental and target characteristics identified in step 1. Concurrently, it would be desirable to assess, wherever feasible, benefits of developing and introducing new componentry specifically suited to realizing optimal resonance detection, defining where and how such new componentry would advance this technique if one were not bound by the quick reaction goal to exploit as much as possible COTS and MOTS hardware and tools.

Step 3. Develop, Manufacture, and Deploy an Operational System

Produce an initial operational system; integrate the system with existing military and law enforcement infrastructure; train users in the employment of deployed technology; continue to expand the signature database before and during field use. Development time through initial operational capability could be 18–24 months for an initial operational system based largely on maximum use of COTS/MOTS tools.

A definitive timeline should also be established in step 3, if substantial new componentry is determined to be accessible to development, for an improved resonance detector product having desirable capabilities significantly improved over the initial operational system. The timeline for such an enhanced or product-improvement resonance detector might be expected to be perhaps 24–36 months after availability of the initial operational system. This improved system development would be undertaken only if it turned out that an initial system based on COTS/MOTS tools had major desirable growth potential with new componentry.